The LATEST GLOBAL VOL.2
Show Flat Collection

全球最新样板房设计大赏(下)

深圳市创扬文化传播有限公司 编

中国林业出版社

The LATEST GLOBAL Show Flat Collection
全球最新样板房设计大赏
CONTENTS 目 录

004	公园大地36栋806	Room 806, Building 36 of Land Park
010	公园大地33栋03户型	Room 03, Building 33 of Land Park
016	公园大地32栋1901	Room 1901, Building 32 of Park Land
022	公园大地26栋2001	Room 2001, Building 26 of Park Land
030	怀特街35号	35 White Street
034	透明的阁楼	Transparent Loft
038	曼谷Thong Lor私人住宅	Private Residence, Thong Lor, BANGKOK
042	汤普森住宅销售中心	Thompson Residences Sales Centre
046	低调华丽古典美学	Understated Gorgeous Classical Aesthetics
052	庆泽园陈公馆	Qingzeyuan Chen's Residence
058	福清融侨国际公馆	Fuiqing Rongqiao International Mansion
062	西岸故事样板房1	Western Coast Show Flat (1st)
070	西岸故事样板房2	Western Coast Story Show Flat (2nd)
076	西岸故事样板房3	Western Coast Story Show Flat (3rd)
082	西岸故事样板房4	Western Coast Story Show Flat (4th)
088	汇景新城	Hui Jing New City
094	星光华庭1栋601	Room 601, Building 1, Xingguang Huating
102	中源名都-新古典欧式	ZHONGYUAN MINGDU-NEO CLASSICAL STYLE
108	中源名都 ARTDECO	Zhongyuan Mingdu ARTDECO
114	中源名苑 新古典	ZHONGYUAN MING YUAN NEW CLASSICAL
120	居住主题公园	Theme Living Park
126	辽宁宗骏地产样板房	Liaoning Zongjun Show Flat
130	烟台西海岸多层样板间	Duplex Show Flat of Western Coast in Yantai
136	万城华府	Wan Cheng Hua Fu
144	雅居乐花园	Elegant and Pleasant Garden Residence
152	融信宽域	RongXin KuanYu
158	江岸公馆-现代时尚	Mansion by Riverside-Modern Fashion
164	江岸公馆-绿中海	Jiang'an Mansion-Green Sea
170	阿姆斯特朗大道	Armstrong Ave

174	巴伊亚住宅	Bahia House
180	**KKC**	**KKC**
186	**KRE**	**KRE**
192	汇璟花园	Huijing Garden
198	三盛巴厘岛	Sansheng Bali Island
202	缇香水岸	Luxury Residence by the Riverside
208	中央美苑	Central Beautiful Garden
214	黑色流金	Noble Black
222	芳House	Fang House
228	福州大儒世家	Fuzhou Confucian Family
232	公园道一号	Gongyuandao No.1
236	东方润园	Dongfang Runyuan
242	江南水都美域（一）	Beautiful Residence in Jiangnan(1st)
246	东尚观湖样板间—精致生活	Dongshang Guanhu Show Flat-Elegant Life
250	新东尚样板间–绅士主张	New Dongshang Show Flat-Gentlemen's Choice
254	金地国际花园	Jindi International Garden
258	福熙大道	FuXi Dadao
264	泛海国际	Fanhai International
270	丽景天成	Natural Beauty
274	色韵	Color Charm
278	海语西湾	Haiyu Xiwan
282	卓越维港	Zhuoyue Weigang
288	依山郡	House besides Hills
298	香瑞园霍公馆别墅	Xiangrui Garden Huo's Villa
304	蓝郡海蓝G型复式豪宅	Sky-blue G-type Duplex Villa
312	钱隆首府	Qianpu Manor
318	万科金色城品	Vanke Jinse Chengpin
324	星湖丽景复式样板房	Starry Lake Scenery Duplex Show Flat

Room 806, Building 36 of Land Park

公园大地36栋806

设计公司：戴维斯(国际)设计及顾问有限公司　项目地点：深圳
建筑面积：123平方米　主要材料：玻璃、茶镜、水晶珠帘、地毯

This project features with unique fragrance. Strict edge decorations and smooth surface is matched up with delicate crystals to create an attractive and intoxicating space, which reveals the charm of the new age. The entire space contains several coexisting units which include open sitting room and multifunctional spaces, the living room, dining room, kitchen, bar and study can all of them can be united as one.

In the master bedroom, the bed background wall is made of golden materials and decorated with mirror surface, which are full of modern flavor. The words written on the wall "Always with joy" shows the owner's life attitude. Cozy, comfortable, leisure and natural, with those qualities, this is just the home the owner pursues for.

　　本案以创新独特的琳琅馥郁为特色。硬朗的边饰与光滑的表面，再结合精致的水晶塑造出动人的迷醉空间，并流溢出新世纪魅力。整个空间设计构造出一个共栖式的生活单元，它拥有开放式起居及多功能空间，客厅、餐厅、厨房、酒吧、书房均可合为一体。

　　主卧床背景墙由金色面料和镜面装饰，富有现代感，上面的那句"Always with joy"，点明了业主的生活态度。温馨、舒适、随意、自然，这正是业主追求的居家空间。

Room 03, Building 33 of Land Park

公园大地33栋03户型

设计公司：戴维斯(国际)设计及顾问有限公司　项目地点：深圳
建筑面积：167平方米　主要材料：皮革座椅，包豪斯地毯，水晶珠帘

This project adopts elegant and noble design style, which shows the taste and dignity of luxurious house. The golden color furniture and lightings in the warm color dominated space create a dignified scene. Besides, the golden shining crystal bead curtains, the soft leather chairs as well as the Bauhaus carpet imbue the space with noble yet modest charm.

The mirror surface covered ceiling is matched with crystal droplights, which reflect golden shinning artistic effect and bring visual extendibility. The huge French window not only permits abundant lights into the space, but also allows people to view the beautiful scenery outside.

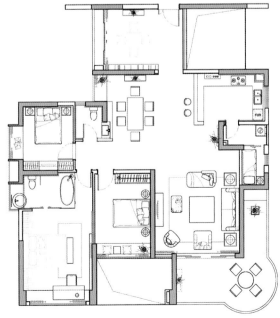

　　本案采用典雅贵气的设计风格，突显豪宅应有的格调和气派。在这暖色中配上金色调的家私及灯具营造出贵气不凡的空间场景。加上金灿灿的水晶珠帘，触感柔软的皮革座椅配上包豪斯地毯，令空间散发出高贵含蓄的醉人韵味。

　　镜面铺就的天花，加上水晶吊灯的装饰，折射出金光灿灿的艺术效果，带来极致的视觉张力。巨大的落地窗，既能给室内空间带来充足的光线，也能让人们在室内一览屋外的美景。

03 Room 1901, Building 32 of Park Land

公园大地32栋1901

设计公司：戴维斯(国际)设计及顾问有限公司　项目地点：深圳
建筑面积：200平方米　主要材料：石材、马赛克、大理石、玻璃、茶镜

This project is modern, gorgeous and classical, as the design isn't limited to fixed rules, clear visual effect creates incomparable striking style of the interior. The easy and various colors reveal luxurious style as well as infinite charm. Roughness and smoothness, matt and gloss, deepness and brightness...all kinds of materials and colors are perfectly combined to form a delicate balance, creating a harmonious space.

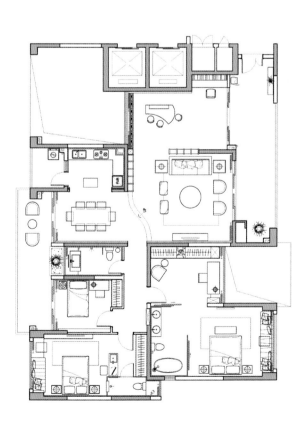

本案现代、华贵而经典，设计不限于条框，清晰的视觉效果带出室内无可比拟的强烈风格。轻盈飘逸富有缤纷的色彩，散发出奢华的风格和无限的魅力。粗犷与平滑，哑面与光面，深沉与明亮……各种物料、色彩相互融合，展现出细腻的平衡，营造了一个和谐的室内空间。

04 Room 2001, Building 26 of Park Land

公园大地26栋2001

设计公司：戴维斯(国际)设计及顾问有限公司　项目地点：深圳
建筑面积：202平方米　主要材料：玻璃、大理石、软包、茶镜

An exquisite and elegant living space is the ideal home of urban aristocrats. In this project, by careful designing of the details and the reasonable use of glass, tawny mirror and marble, an elegant home interior is created, besides, one can feel luxurious and modern flavor.

Luxurious pile carpet contrasts with the transparent and splendid droplights, demonstrating the magnificence of the living room. The dining room is matched with open style kitchen, which extends the visual effect of the dining area, and in the meantime, it promotes luxurious and elegant vibe to let people indulge in the warm dining atmosphere.

　　精致典雅的生活空间，是都市新贵的理想之家。本案通过对细节的追求和对玻璃、茶镜大理石等材料的合理运用，成功地打造出一个雅致的居家空间，略带奢靡摩登的味道顿时跃然眼前。

　　华丽的绒毛地毯与晶莹的华丽吊灯形成对比，烘托出整个客厅的华美。餐厅搭配开放式的厨房，伸展了就餐区的视觉效果，同时造就了豪华高雅的气氛，使人沉浸于舒适的就餐环境之中。

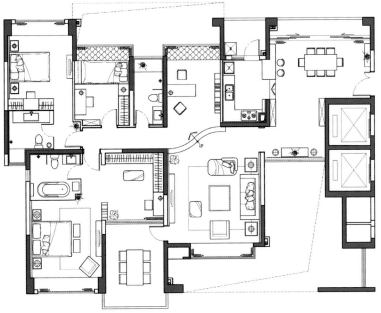

05 35 White Street

怀特街35号

设计公司：Andre Kikoski Architect　项目地点：美国纽约
建筑面积：176平方米　摄影师：Peter Aaron

This loft in the heart of Tribeca was the painting studio of abstract expressionist Barnett Newman. Daylight from one hundred feet of windows fills the space. Our intention to preserve and enhance the essential character of this historic loft -- while crafting within it a domestic space for a family with a young child – relied on a sense of openness and light. Operable panels of prismatic glass configure to create either dramatic openness or total privacy. Custom mahogany and bronze panels enhance the natural breadth and depth of the loft. A state-of-the-art cook's kitchen and French limestone bathrooms make this space equally suited for dramatic entertaining and intimate daily living.

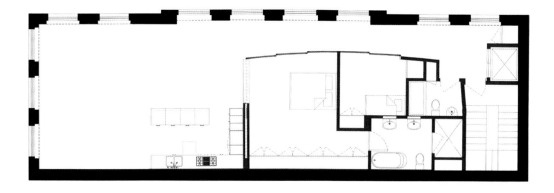

　　这是一个位于曼哈顿特里贝克中心区的小阁楼，曾经是一个抽象表现主义画家Barnett Newman的画室。日光从离窗户30米的地方照射进来。我们的目的是在里面为一个小朋友精心雕琢一个属于自己的宽敞且光线充足的家，同时保护和提高这座历史性阁楼的本质。可操作的嵌板和五光十色的玻璃打造了一个充满戏剧性且宽敞或者绝对隐秘的空间。定制的红木盒古铜色的嵌板加强了其自然的宽度和阁楼的高度。顶级水准的厨房和法式石灰岩浴室使这个空间同样也适合娱乐和亲密的私生活。

06 Transparent Loft

透明的阁楼

建筑师：Olson Kundig Architects 领导设计：Jim Olson 合作团队：Jim Olson（设计理念）；Valerie Wersinger（项目经理）；Janice Wettstone（设计师助理）
室内设计：Ted Tuttle 项目地点：西雅图 建筑面积：242.5 平方米

This eighteenth-floor condominium in downtown Seattle carries the idea of transparency to its logical extreme. The goal was to improve the boxy proportions of a speculative apartment, giving it the openness of a converted loft. The kitchen and master bathroom are enclosed with walls of glass to match the expanses of glazing on two of the exterior facades and around the recessed terrace. The elevation of the apartment assures privacy, as does the separation of public and private areas with a wall and sliding screen.

The interior design is minimal in material and palette. A polished black floor sets off the glass and white walls and is warmed by wood tables, paneling, and the soft tones of upholstered seating. In the bathroom, mirrors mounted on the glass walls feel suspended in space; the same illusion is presented by metal fittings and sleek cabinets in the kitchen.

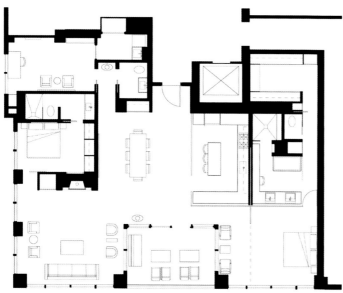

　　这是一个18层的公寓，坐落在西雅图的市中心，它使透明设计发挥到了极致。我们的目的是提高异型公寓四方的比例，给它提供一个开阔的、可以变幻的阁楼。厨房和主浴室是用玻璃墙打造的封闭式区域，这是为了与室内大片的玻璃窗和休息阳台相匹配。公寓的高度保证了其私密性，就像是分离的公共空间和一个带有墙壁和推拉门的私人空间一样。

　　室内的设计使用了最少的材料和调色板。光滑的黑色地板衬托出了玻璃与白墙，并且也因为木质桌、嵌板和暖色调的软垫桌而变得温暖了。在浴室，镜面镶嵌在玻璃墙上，感觉像悬空的。厨房里光滑的橱柜、五金装饰也呈出了同样的幻觉。

Private Residence, Thong Lor, BANGKOK

曼谷Thong Lor私人住宅

设计公司：MONOCHROME INC　项目地点：曼谷　建筑面积：280 平方米

The project was the renovation of a villa in the Thong Lor precinct of the Bangkok, that was originally constructed in 1969. The design of the villa was a combination of American and Thai influences, and it was built in the split level style that was very popular at the time. The interior featured dark timber doors, detailing and architraves throughout, and consequently, overall was quite a dark house. The renovation included a new interior design concept based on a monochromatic theme. The dark timbers, wall and ceiling surfaces were replaced with a bright and fresh paint scheme in white, with the selected furnishings in a combination of white and black with chrome detailing. A beautiful bright red was chosen as the single accent color to compliment the black and white design, and this was reflected in artworks and other accessories throughout the villa.

此项目是为曼谷Thong Lor区的一座建于1969年的别墅翻新。别墅的设计受到美式和泰式的影响,并且结合了当时流行的各种风格。深色的木门、点缀物和柱顶过梁贯穿整个室内,因此,整体上形成一个暗色调的房屋。翻修时新的设计理念便是以单色调主题为基础。暗色的木材、墙壁以及天花板的表面都新上了一层明亮的白色涂漆,并且挑选了黑白的家具、陈设品和铬合金点缀物与之搭配。美丽的鲜红色被选作与黑白设计搭配的单色调,这从整栋别墅内的艺术品和其他配件可以反映出来。

Thompson Residences Sales Centre

汤普森住宅销售中心

设计师：BURDIFILEK 设计团队：Diego Burdi、Paul Filek、Amy Chan、Spencer Lui、Thomas Moore、Anna Nomerovsky、Helen Chen、Jacky Kwong、Samer Shaath、Sarah Gatenby
项目地点：加拿大多伦多 建筑面积：约297平方米 摄影师：A Frame公司Ben Rahn

The space adopts no-door design, the areas are all semi-open or open style. In the interior, the wall, floor, ceiling and furniture ornaments as well as lightings are all featured with minimalist appearance, pure material and delicate artwork. The design for the furniture pays much to its function, with simple flowing lines and contrast colors. With unique design, the simple lines or even the creative ornaments can be decoration of the space.

The bedroom also adopts minimalist design style. On the bed background wall, there are two decorative paintings, and on the opposite, there is a floor ceiling. There's no redundant decorations in the space, all of them have functions and practicability.

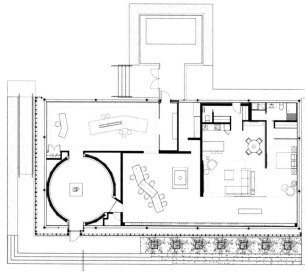

　　空间采用无门的设计，均是开放或半开放的区域。室内墙面、地面、顶棚以及家具陈设乃至灯具器皿等均以简洁的造型、纯洁的质地、精细的工艺为特征。家具的设计突出强调功能性，线条简约流畅，色彩对比强烈。简单的线条，独特的设计，甚至是极富创意和个性的饰品都可以成为点缀装饰空间的一员。

　　卧室也是采用极简的设计，床背景墙上有两幅装饰画，对面墙放置了一面全身镜。空间内没有多余的装饰，一切都是以实用为主。

09 Understated Gorgeous Classical Aesthetics

低调华丽古典美学

设计师：白瑞翔　设计公司：关御室内装修有限公司　项目地点：台中
建筑面积：363平方米　装饰材料：樱桃木皮、帝诺大理石、深金峰大理石、烤漆玻璃、段造铜、琉璃、裱布、大面茶镜、透光石、壁纸

In this project, the designer focused on the combination of noble, elegant and delicate Neo-classical style and colorful, non-traditional Baroque style. On the base of his bold choice of the materials and color match, with imagination and creative ideas, the space becomes a well matched living environment.

The villa includes three gardens on three sides, and inside the villa, the first floor serves as public area, the second floor and above are private spaces. The interior is very broad and deep. For the public space, the two living rooms are imbued with delicate and colorful baroque flavor. Irregular parquet Tino marble creates stereotypic layers of the TV background wall. The ceiling to floor mirror reflects the rich sceneries of the space, and the bright colors are so fascinating that people would feel like wandering in a starry castle happily.

　　本案中，设计师强调的是高贵典雅、雕刻精细的新古典风格与色彩丰富、颠覆传统的巴洛克风格的结合。以大胆的素材选用和色彩搭配为主轴，丰富的想象力让整体空间的结合增添许多创意并创造出混搭的居住环境。

　　别墅建筑拥有三面花园，内部格局以一楼为公共厅区，二楼以上则为私密空间。室内空间宽广、纵长景深，公共空间以细节繁复、色彩鲜艳的巴洛克风格贯穿两厅，不规则的帝诺大理石拼贴出电视墙的立体层次。落地镜饰映射空间的富饶景象，饱和的色彩令人沉醉迷恋，像是飘飘然漫步在星光璀璨的城堡里。

10 Qingzeyuan Chen's Residence

庆泽园陈公馆

设计师：黄俊盈　设计公司：梵古设计开发中心　项目地点：台湾台北
建筑面积：138平方米　主要材料：阿拉斯加香杉、玻璃、石材

In this case, besides visual and tactile senses, the designer incorporated olfaction, utilizing Alaska cedar as the olfaction source of the space. The hallway, and end tables of various sizes in the living room, dining table and chairs, the Bar counter and chair, the bedside cabinet in the master bedroom as well as the hand cut round chair, they all give off the Pythoncidere fragrance of cedar, relieving the residence's pressure.

The space focuses on leisure, using simple lines with clear angles, without redundant wall modeling; exquisite furniture and wall ornaments are selected to create an attractive tactile space. The design for children's room is more interesting; the red Chinese figured cloth, the rocking chair developed from dinging chair, the bare pine tree desk in the girl's room; the skillful combination of basketball hoop, desk and chairs in the boys' room, all these make the space more lively and charming.

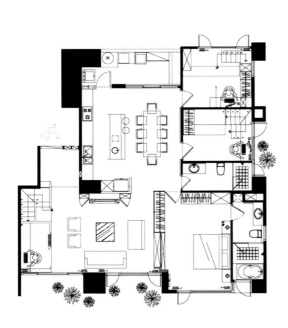

　　在本案中，设计师尝试在除了视觉、触觉外加入嗅觉，利用阿拉斯加香杉作为空间嗅觉来源。从玄关端景到客厅大小茶几、餐桌椅及Bar台椅，最后到主卧室床头柜、窗台手工刨切圆单椅，利用香杉散发出来的芬多精香味，舒缓居住者的身心压力。

　　空间定位为都会休闲，利用简单的线条切割，不强调太多的壁面造型，选用精致的家具、壁饰等，烘托出空间的质感。较有趣味性的是小孩房间的设计，女孩房的中国大红花布装饰、由餐椅改良设计的摇椅、不上漆的松木书桌；男孩房中篮筐、书桌与楼梯的巧妙结合，这些让空间显得更加活泼灵动。

Fuiqing Rongqiao International Mansion

福清融侨国际公馆

设计师：陈明晨、沈江华、陈墩华　设计公司：鼎汉唐（福州）设计机构
项目地点：福建福清　建筑面积：135平方米
主要材料：金刚板、蒙托漆、墙纸、葡萄牙软木、进口仿古砖

The entire interior is overally schemed in coffee color, transitioned with beige and pale tinge to make sober yet harmonious ambience. Furnishing is well-mixed, as the deep colored ones showing personality and taste of the designer throughout its poise. As for the lighting, point, line and surface sources are organically combined to highlight the hierachy and dimension. The use of materials like the cork with conspicuous natural texture, the mengtuo paint, the iron plate etc. enhances the affinity while showing the minimalism of the space.

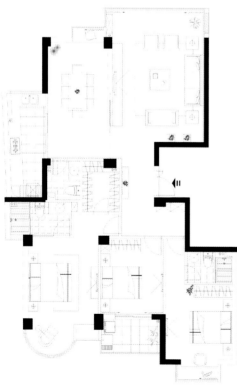

本案的设计采用的是简约风格，室内整体色调以咖啡色为主题色，搭配米色及浅咖色作为过渡，使整个空间更具沉稳的同时又不失和谐。在家具配套上，采用了混搭的手法，深色家具的使用在沉稳中体现个性与品味。灯光上，采用点光源、面光源、线光源的有机结合，使空间更有层次，立体感更强。材质上使用自然肌理质感较强的软木、蒙托漆、金刚板等，使室内空间在体现简约感的同时更具亲和力。

Western Coast Show Flat (1st)

西岸故事样板房（一）

设计师：刘威　设计公司：武汉刘威室内设计有限公司　项目地点：武汉市洪山区
建筑面积：73平方米　主要材料：抛光砖等　摄影师：吴辉

This is a minimalist and delicate show flat. Though the space is not large, it's imbued with the simplicity and neatness of Hongkong style.

The living room is dominated with cold color, the white floor tiles, light grey sofa and the randomly placed delicate tea set and magazines endow strong vitality to the space. In order to match the cold color tone of the living room, the designer selected wooden color TV background wall to balance the grey and white in the living room, attaining color harmony.

The design for the room extends minimalist style. Storage boxes are orderly placed in the cupboard, some higher, some lower, clean and neat. Matched with distinctive ornaments such as black candle stick, fish model, then life flavor is produced.

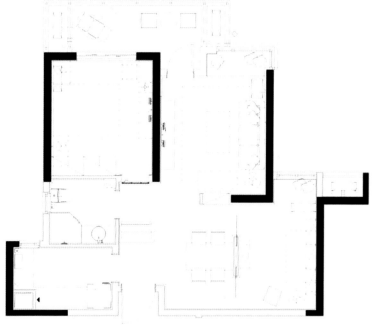

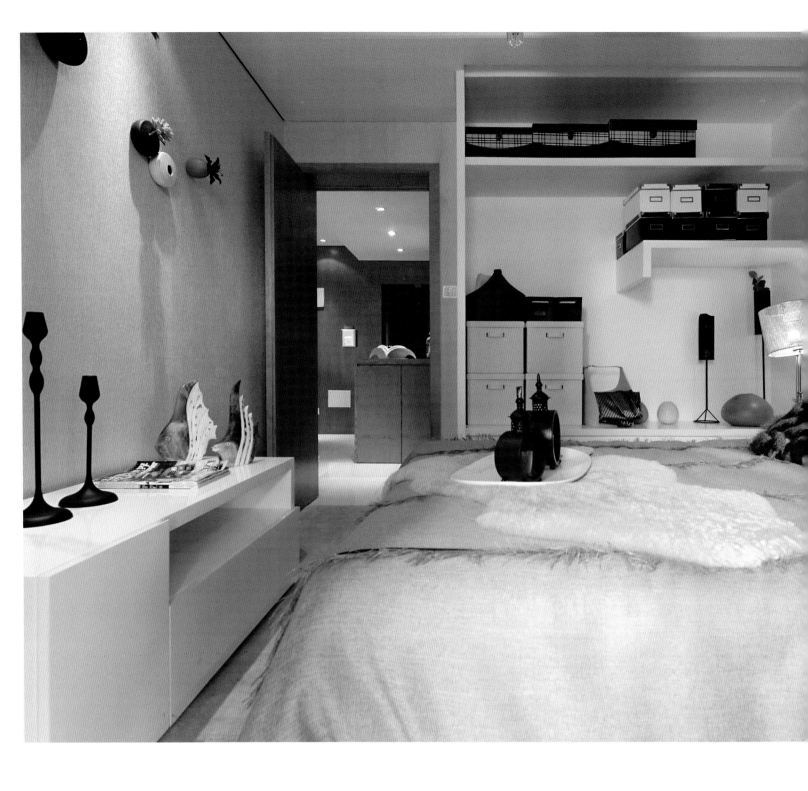

　　这是一个简约细致的样板间。空间不大,却充满了港式风格简单、利落的味道。

　　客厅的主色调为冷色调,白色的地砖,浅灰色的沙发,茶几上随意摆放的别致茶具和杂志,浓浓的生活气息跃然眼前。设计师为了调和客厅的冷色调,精心选用木色的电视背景墙,平衡了客厅的灰白色调,达到了颜色上的和谐。

　　房间的设计延续了简约的风格,储物箱被错落有致地安放在衣柜上,干净而整齐。再配以独特的小饰物,如黑色烛台、海鱼摆件,生活的味道也就出来了。

Western Coast Story Show Flat (2nd)

西岸故事样板房（二）

设计师：刘威 设计公司：武汉刘威室内设计有限公司 项目地点：武汉市洪山区
建筑面积：87平方米 主要材料：木地板、沙发、装饰画 摄影：吴辉

A romantic house imbued with the flavor of fairy tales must be every girl's ideal home. This project is about a romantic show flat, which is based on bright white, in the warm light, people would feel like being in a kingdom of fairy tales.

White sofa, white wooden floor, added with white wall, such a match may give the impression of dullness, but with the designer's skillful design, by pink and emerald sofa cushion, simple yet exquisite background painting of flowers, the entire living room is turned into a lady from a scholarly family, graceful and elegant.

The design for the door of the bathroom gives the most pleasant surprise. Looking from afar, it's a painting of engraved roses and butterflies, but upon close watch, one would be surprised to find that it's a hidden door. What a wonderful design! And one can't help admiring the designer's carefulness.

　　浪漫而又充满童话色彩的房子，相信是每个女孩子的梦想之家。本案是一个充满浪漫气息的样板间，整个房子的主色调是亮丽的白色，在温暖灯光的映照下，仿佛让人身处童话王国。

　　白色沙发，白色木地板，加上白色墙面，这样的配搭也许容易让人感到单调，但在设计师的巧妙转化下，配以粉色、翠绿的沙发靠垫，简单精致的花卉背景画，整个客厅就像一位书香门第的淑女，既高贵又有气质。

　　但最让人感到惊喜的是卫浴门的设计。远看是一扇刻满玫瑰和蝴蝶的装饰画，细细一看，才发现其中的奥妙，那竟是一扇隐藏得很好的门。这样贴心和巧妙的设计，不得不赞叹设计师的细心。

Western Coast Story Show Flat (3ʳᵈ)

西岸故事样板房（三）

设计师：刘威　设计公司：武汉刘威室内设计有限公司　项目地点：武汉市洪山区
建筑面积：87平方米　主要材料：仿古砖、木制家具　摄影师：吴辉

Opening the door, one would feel like coming into a blue sea. The ancient floor tiles, vintage wooden furniture, casually placed delicate ornaments imbue the house with childhood's pleasure, and in the meantime, it's endowed with the trace of the passing times.

The living room is connected with the dining room, so that the space becomes spacious, even with many lovely ornaments, it won't appear crowded. The white droplights, giraffe ornament and blue candlestick are all with strong Mediterranean flavor.

At that moment, even you're not in Aegean Sea, you could feel its romantic flavor.

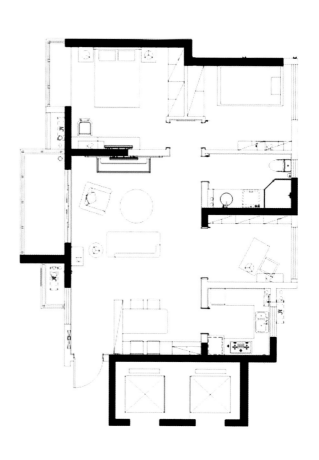

推开门，仿佛进入一片蓝色的海洋。复古的地砖，防旧的木制家具，随意摆放的别致小饰物，让这个房子充满童趣，同时也流淌着岁月时光的痕迹。

客厅和餐厅连在一起，显得空间很大，即使搁置很多可爱的小饰物，也不会显得拥挤。白色吊灯、长颈鹿摆件、蓝色烛台，都充满了浓郁的地中海风情。

此时此刻，即使不在爱琴海，也能感受到爱琴海的浪漫风情。

15 Western Coast Story Show Flat (4th)

西岸故事样板房（四）

设计师：刘威　设计公司：武汉刘威室内设计有限公司　项目地点：武汉市洪山区
建筑面积：93平方米　主要材料：木地板、地毯、壁炉　摄影师：吴辉

This is a pleasant house without any artificial design trace, all is natural and casual, the layout of the furniture and the color match are harmonious.

The living room and dining room are both covered by wooden floor, the wood color is matched with light brown sofa, and there's a beautiful countryside painting behind. In the front, there's a fireplace and a pile of firewood, full of natural flavor.

The most special one is the design for the Children's room. The background is decorated with blue stripes wallpaper, and the bed is with blue and green stripes. The blue curtain, white sailing boat modeling, the refreshing match make people feel like being near to the sea.

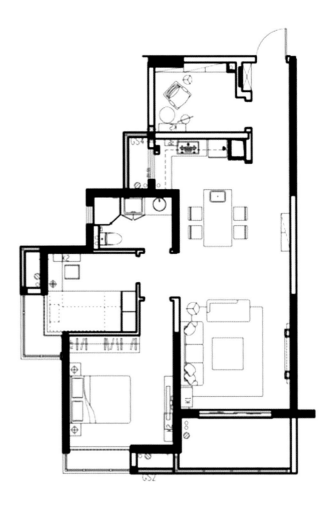

　　这是一个令人赏心悦目的房子，没有丝毫刻意的设计痕迹，一切都很自然随意，家具的摆设以及颜色的搭配都很协调。

　　客厅和餐厅都采用木地板铺就，古朴的木质颜色，搭配浅棕色的沙发，背后是一幅美丽的田园风光图。正对面有一个壁炉，还堆放了一堆柴火，充满了自然气息。

　　最为特别的是儿童房的设计。用蓝色条纹的墙纸作为背景装饰，蓝绿条纹的床铺、海蓝色窗帘、白色的帆船模型，清新的搭配，让人仿佛身处海边。

Hui Jing New City

汇景新城

设计师：George Yabu、Glenn Pushelberg 设计公司：YABU 建筑面积：约470平方米
摄影师：陈启明

The design of this project combines with some antique sense of richness, modern luxury and elegance. The easy and generous spiral staircase, splendid droplight and beautiful bookcase reflect the owner's romantic and creative spirit. The everlasting elegance and beauty are revealed by abundant materials, such as: natural marble, platinum color wood and slivery metal, with beautiful scenery from outside the window, surpassing the past and future while revealing the ambitions and pursuits of modern people.

Every light and every chair is customized made, in such a highly industrialized period, the custom-made light and chair maybe extravagant, but they could become the classic. Extravagance not always means flaunt and show-off, it means to find the peace in balance, it is cozy, exquisite and elegant. This depends on carefully carving by inspiration and intuition and creates a beautiful balance.

　　本案的设计中融合了古典的富裕感与现代的豪华与精致。从容气派的旋转阶梯、灿烂的吊灯和美观的书架，反映了主人既富有浪漫色彩，又拥有创造精神。永恒的优雅和美感经由丰富的材料如天然大理石、白金色的木材和银色金属等呈现出来，配合着窗外的美景，跨越了过去与未来，描摹出当代人的雄心和追求。

　　每盏灯、每张餐椅都是专门定制，在泛滥着工业制品的时代，这种定制也许是奢侈的，但它更有可能成为经典。奢华就是在物体的平衡中寻找宁静，并不总意味着招摇和炫耀，它可以既舒适精巧，又流露品位。这是依靠灵感与直觉精心雕琢，创造出的一种优美的平衡。

Room 601, Building 1, Xingguang Huating

星光华庭1栋601

设计师：胡媛媛　设计公司：汕头市丽景装饰设计有限公司　项目地点：汕头
建筑面积：131平方米　主要材料：木地板、墙纸、乳胶漆、地砖等

The design for this project emphasizes the independence of the space, each space has its own distinctive function. The sofa and carpets are European style, which enclose this elegant and living room. The sofa background wall is decorated with golden-patterned red wallpaper, which echoes with the red carpet, forming a harmonious match.

As for the layout, the designer pays much attention to the independence, function and comfort of the space. The layout and decoration of each area is delicate, with full consideration to the owner's needs. The designer has successfully created a luxurious and easy home interior for the owner.

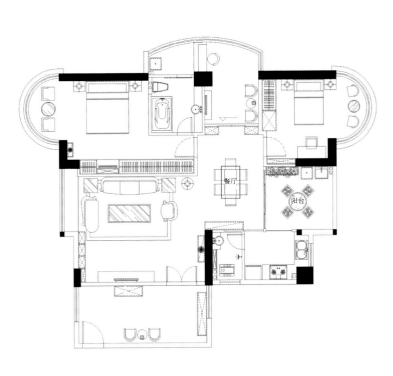

　　本案在设计时强调了空间的独立性，各个空间有着明确的功能划分。沙发、地毯、窗帘，全部都是欧式风格，围合成这个雅致、华丽的客厅。沙发背景墙选用红色带有金色花纹的墙纸，与红色的地毯相呼应，给人以和谐舒适的感觉。

　　设计师在平面布局上，讲究独立性，重视空间的功能性和舒适度。每个区域的布局和造型都非常精致，充分考虑到居住的舒适度，为户主建造了一个豪华、自在的居所。

78 ZHONGYUAN MINGDU-NEO CLASSICAL STYLE

中源名都-新古典欧式

设计师：D组/黄炽烽、欧敏华　陈设与选材：肖卉、陈洁玲、区婷婷
设计公司：J2-STUDIO/厚华顾问设计有限公司　项目地点：广东省肇庆市
建筑面积：250 平方米　主要材料：灰钢、墙纸、灰镜、蚀花玻璃

This project adopts Neo-classical style, with black brown as the dominated color, breaking away from the traditional design of white and beige as the dominated colors. Black is boldly used for the furniture, lighting as well as wallpaper. In this project, the major materials include grey steel, wallpaper, grey mirror and flower-patterned glass; another highlight is the wide use of leather, which is used from the living room to the bathroom. Besides the dominated Neo-classical style, some modern design languages are incorporated, such as the TV background wall, the bookcase in the study, the background wall in the master bedroom, they are all with stylish flavor.

　　本案采用新古典风格，整体风格以黑褐色调为主，打破传统的白色、米黄色古典设计方向，包括家具、灯具、墙纸等都大胆地使用了黑色。本案主要材料为灰钢、墙纸、灰镜、蚀花玻璃，另外一个亮点是运用了比较多皮质材料，从客厅到卫生间都有运用。而在新古典为主的风格外，也注入了很多现代构成的设计语言，例如电视背景墙、书房书柜、主人房背景墙等都充满了时尚现代的质感。

Zhongyuan Mingdu ARTDECO

中源名都 ARTDECO

设计师： D组/黄炽烽、欧敏华　陈设与选材：肖卉、陈洁玲、区婷婷
设计公司： J2-STUDIO/厚华顾问设计有限公司　项目地点：广东省肇庆市
建筑面积： 150平方米　主要材料：灰镜、灰钢、紫色软包造型等材料

This project demonstrates modern ARTDECO life style. Various flower modeling of different materials are used in different spaces to show the diverse themes, and here, people would feel the tender beauty of female.

The materials used in this space include wallpaper of dark color large flowers, glass with printed flowers, grey mirror and grey steel, purple soft furnishings, and the wide use of wall paper with large flower patterns imbue the space with silver grey tone, it's matched up with modest mysterious purple to connect the major spaces such as the living room, dining room corridor and the master bedroom.

The water wave subject is a highlight of this project. Behind the bed, the purple sandblasting glass modeling and the carpet form beautiful lines, demonstrating a noble and luxury living conscious, and they also bring visual surprise and pleasure.

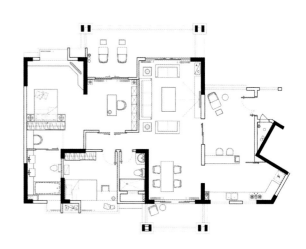

本案主要体现了一种现代ARTDECO的生活风格。各种花的造型在各个空间通过不同的材质表现形成了不同的主题，让人感受当中一种仿似女性般柔美的情怀。

空间主要用了暗纹大花墙纸、印花图案玻璃、灰镜、灰钢、紫色软包造型等材料，通体大花墙纸的使用形成了整个空间的银灰色基调，配合低调神秘的紫色调，连通了客厅、餐厅过道和主人房等各个主要空间。

主人房的水纹主题是整个案例的一个亮点，床背的喷砂紫玻璃造型与地毯形成的曲线美态，显示了一种高贵奢华的生活意识，在视觉上也充满了动感与惊喜。

20 ZHONGYUAN MING YUAN NEW CLASSICAL

中源名苑 新古典

设计师： D组/黄炽烽、欧敏华　陈设与选材：肖卉、陈洁玲、区婷婷
设计公司：J2-STUDIO/厚华顾问设计有限公司
项目地点：广东省肇庆市　建筑面积：220平方米　主要材料：雪茄木、马皮、夹丝玻璃

This project adopts Neo-classism style. The space is divided into three floors. The first floor functions as the sitting room, and the ground floor serves as an entertaining area, the second floor underground contains multimedia room, reception room, bar, sports room and resting room. The entire hard furnishing has little trace of classical elements, but the soft furnishings all adopt Neo-classical furniture and ornaments. Brown sets the major color tone, and the main materials including wire glass, cigar wood, and horsehide soft furnishings are matched one another, showing the steadiness and magnificence of the space.

　　本案为新古典风格，空间分为3层，一层为主要起居功能，负一层为娱乐设施功能，负二层为停车场。其中负二层包含功能有影音室、接待室、酒吧、运动室、休息室。整套风格在硬装上没有保留过多的古典风格元素，而注重在软装上采用统一的新古典风格的家具和饰品。色彩搭配上主要以褐色调为主，材料上运用了夹丝玻璃、雪茄木、马皮软包等的搭配，表现出空间较沉稳和大气的性格。

21 Theme Living Park

居住主题公园

设计师：陈锐峰　设计公司：福州中和设计事务所　项目地点：福州
建筑面积：300平方米　主要材料：蜜蜂瓷砖、阿曼米黄大理石、德国壁纸、文行灯饰

The designer makes solemn and elegant door partition merge into the space, giving people a sense of strong astonishment. The metal edge makes the mirror in the middle become the focus of the space and adds the generous quality into the space.

The square living room with luxurious decoration makes the entire space appear distinguished and generous. The TV setting wall is made up of three arched doors which embody the strong flavor of European style and the designer delicately deals with the relationship of the doors in the living room. The marble floor, the antique furniture and the crystal droplight decoration of living room make the classic feeling completely revealed in this generous and elegant European design style.

The living room and dining room are connected as one and they form an open layout. Even you are having dinner, you could see the living room, thus, it has much fun for the owner to have dinner there.

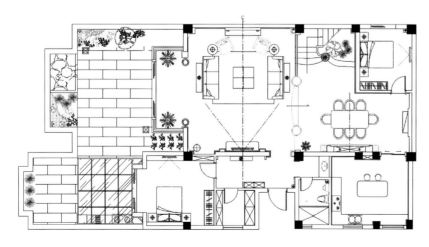

　　融合了庄重和优雅双重气质的回字形门屏，在一进门就给人以强烈的震慑力。金属造型的镶边让中间的镜面轻易地成为空间的焦点，多了几分王者霸气。

　　方正的客厅，配上豪华的装饰，整个空间感觉高贵大气。电视背景墙由3个拱门组成，体现了浓浓的欧式情怀，也巧妙地处理了客厅中几扇门之间的关系。由大理石地板、仿古家具及水晶吊灯装饰的客厅，现代人的古典情怀在这大气雅致的欧式风格中得到满足。

　　客厅、餐厅连为一体，组成开放式的格局。即使在就餐也能了然客厅的一切，增添了业主的用餐乐趣。

22 Liaoning Zongjun Show Flat

辽宁宗骏地产样板房

设计师：王锐　设计公司：沈阳点石（国际）室内设计顾问有限公司
项目地点：辽宁　建筑面积：95平方米　主要材料：纹样墙纸、金镜、实木复合地板、水晶珠帘

This project is featured with Neo-classical style, incorporating modern minimalist with classical elegance. In the living room, the rarely seen leather end table is matched with minimalist leather sofa, and decorated with velvet back cushion, fully showing the classical nobleness. In order to avoid redundant decoration, the designer subtly incorporated minimalist-shaped, classical-flavored TV stand, and instantly the space becomes magnificent.

The master bathroom is a highlight of this project, the washing stand adopts flowing curving outlines and white table-board, appearing minimalist yet generous. Besides, the dressing mirror contains another mirror in the center, which perfectly embodies the fashion and classicalism.

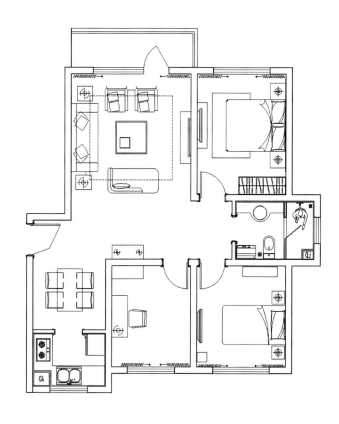

　　本案定位于新古典风格,将现代简约完美地融合于古典高贵之中。客厅用少见的皮质茶几,结合造型简单的皮质沙发,并饰以天鹅绒材质的靠枕,尽显古典高贵。为了避免繁琐之感,设计师巧妙地加入造型简约、略带古典风的电视柜,顿时使空间显得大气。

　　主卫也是本案的一大亮点,洗手台采用流畅的弧形线条,配以白色台面,尽显简约大气。同时梳妆镜采用镜中镜的造型,时尚与古典得以完美体现。

23 Duplex Show Flat of Western Coast in Yanta

烟台西海岸多层样板间

设计师：王锐　设计公司：沈阳一工室内设计事务所
项目地点：烟台　建筑面积：150平方米　主要材料：墙纸、金茶镜、泰柚木

In this project, black, white and grey are used as the dominated color to create a clear and sharp interior. In the space, simple and clear lines are mainly used as the partition or decoration, such as the glass end table, the frame of wine cupboard, the mirror on the washstand and the long bathroom.

In order to relieve the hardness of this space, the designer added some smart elements, such as the shiny ornaments at the turning corner of staircase, the droplights that look like rows of lighted candles, the dancing butterfly on the bedside, those details are well matched with the minimalist square shape, perfecting the interior ambiance.

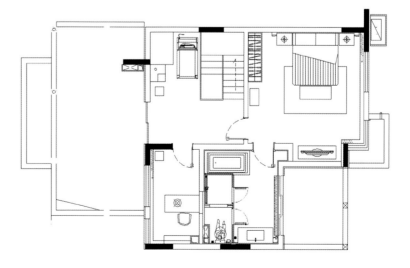

　　本案以黑、白、灰为主色调，打造出一个稍显硬朗的居家环境。空间多以简单明快的线条作为隔断或装饰，如玻璃茶几、储酒柜的造型外框、洗手台上方的化妆镜和长方形的洗浴室。

　　为了缓和空间的硬朗感，设计师给空间注入了一些灵动的因素，如楼梯转角处闪亮的装饰品、餐桌上方仿若一排排点燃的蜡烛的造型吊灯、床头翩翩起舞的蝴蝶，这些小细节与整体的简约方正相结合，更加完美了居住氛围。

24 Wan Cheng Hua Fu

万城华府

设计师： 吴巍　设计公司：北京东易日盛装饰股份有限公司　项目地点：北京
建筑面积：300平方米　主要材料：石材、烤漆镜面、地毯

This is a modern design style case. The entire space is comprised of living room, dining room, kitchen, bedroom, study room and audiovisual studio. The living room, dining room are based on stone materials, baking varnish mirror surface and soft furnishings, interpreting the space air while making the entire space more bright and generous. The TV setting wall is a beautiful night scene, the light is bright and sparkling, giving people strong visual impact. Dark color is widely used, matched with warm light, the entire space becomes bright and dignified, making people feel cozy.

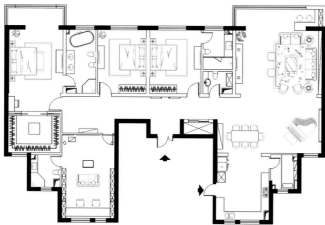

　　本案的整体风格为现代风格。整个空间主要由客厅、餐厅、厨房、卧室、书房、视听等功能空间构成。客厅、餐厅等公共区域以石材、烤漆镜面及软包为主，诠释空间氛围，使整个空间更加敞亮大气。客厅的沙发背景墙是整面的城市夜景图，灯火璀璨，熠熠生辉，给人以强烈的视觉震撼。整体色调选择深色调，配以暖色灯光，让整个空间除了敞亮大气外，更给人一种沉稳而又不失温馨的感觉。

25 Elegant and Pleasant Garden Residence

雅居乐花园

设计师：陈熠　设计公司：北京东易日盛南京分公司　项目地点：南京
建筑面积：180平方米　主要材料：云石、壁纸　摄影：陈熠

This project adopts European Neo-classical style, and with a tint of French flavor, giving people the impression of noble elegance.

The living room and dining room keep their original open style, and form visual harmony by using the same kind of wallpaper. The long corridor forms a sharp contrast with the broad living room.

The entire space is dominated with gentle color, searching balance and change among colors such as gold, champagne and so on. With European style elements and contemporary materials, they also show the owner's life attitude.

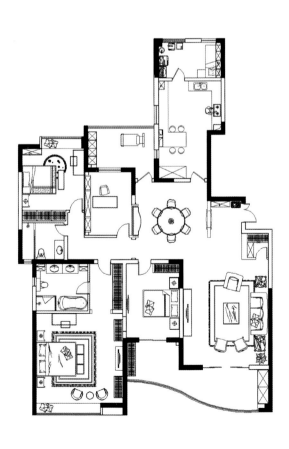

　　本案风格为欧式新古典，其中又掺杂些法式情怀，有高雅华贵之感。

　　客餐厅保留原有开放的空间格局，同时通过采用相同的条纹壁纸在视觉上和谐统一。过道狭长的感觉与开阔的客厅之间形成了明显的对比。

　　整个空间的色彩都是比较柔和的，在金色、香槟色等色系中寻找平衡和变化。造型上有着欧式的元素，材质表现上兼顾了现代感，同时也体现了主人对生活态度上的一种品质。

26 RongXin KuanYu

融信宽域

设计师： 李建林　设计公司：福州三顾室内设计事务所　项目地点：福州金山
建筑面积：110平方米　主要材料：米白色仿古砖、金刚木地板、条纹艺术墙纸

The housewife of this house is very fond of white, and she likes to live in a refreshing and clean living space, so what she need is simple peaceful life instead of complicated luxury.

The design is based on white. The living room and dining room are transparent. The sofa background wall is decorated with three paintings of the same theme; the abstract pictures imbue the space with fashionable vibe. Black quadrate sofa and beige leather sofa are matched with white or stripe-patterned throw pillows and grey curtain, which appear harmonious.

The minimalist design enables people to attain a pure and peaceful heart in the hustling and bustling of city. The complicated and redundant decorations are rejected, so that a neat space is presented.

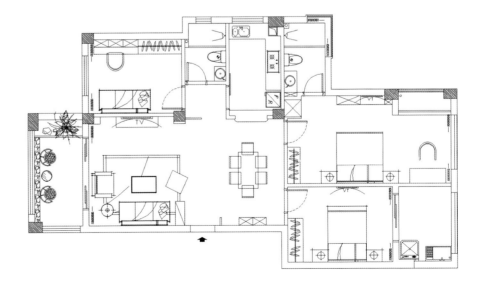

　　本案的女主人对白色情有独钟，喜欢清爽干净的居住空间，不需要喧嚣的奢华，只需要简单安静的生活。

　　设计的基调是白色调，客厅、餐厅采用通透的处理。沙发墙是三幅一体的主题画，抽象的图案，充满了时尚感。黑色的方形沙发、米色的皮质沙发，搭配或白或条纹的抱枕和灰色调的窗帘，整体和谐自然。

　　简约的设计让心灵在喧嚣的都市里得到一种纯净和安宁，摒弃了繁琐和雕琢，显示出井然有序的理性空间。

27 Mansion by Riverside Modern Fashion

江岸公馆－现代时尚

设计师： 毛闯　设计公司：重庆东易日盛责任有限公司　建筑面积： 49.3平方米

The house in this project belongs to small type show flats, which are mainly offered to single white collars or young couples. Comparably speaking, this customer group is those well educated and who have high demands of quality life, so the space should contain noble content.

The project aims to create an exquisite living space, where one could release and show his personality. Mosaic is widely used in this space. Black and white mosaic carpet, mosaic grouping patterns wall plane and wall surface of the kitchen, all these imbue the space with fashionable and modern flavor.

　　本案例属于小户型的示范单位，主要面对单身白领阶层或年轻夫妇。这一类消费者素质相对较高，对品质生活有较高的追求，相应的空间文化内涵也要求更高。

　　本案的设计旨在打造一个精致的生活居所，希望在其中释放自我，展现独立个性。空间多处地方使用马赛克装饰。黑白马赛克的地毯、马赛克组合花纹的壁板和厨房墙面、风格相同的马赛克纹案的窗帘与枕头套……使空间充满了时尚、现代之感。

28 Mansion by Riverside- Green Sea

江岸公馆-绿中海

设计师：毛闯 设计公司：重庆东易日盛责任有限公司 建筑面积：39.2平方米

In this project, the design style is peaceful yet lively. The blue sea and white clouds make people feel so comfortable and relaxed that they forget about themselves. In the living room, log disguised beam is matched with blue and pure white to create a serene and romantic ambiance.

The starfish ornaments and waterweed-shaped bracket light on the sofa background wall, the fireplace with the embellishment of boat and steering wheels on the wall are matched with blue and white carpets, and the living rooms seems to lead people into the mysterious sea. The bedroom is transparent, bright, spacious and casual, the dominated color blue seems to bring the breeze from the seashore, which is so refreshing. The fresh and comfortable bathroom is decorated with blue and white mosaic tiles, which are unique and graceful.

本案的设计充满祥和与生机，蓝蓝的海，白白的云，让人心旷神怡、沉醉其中。客厅使用原木假梁，配合深邃的海蓝色，纯净的白色，打造一个充满宁静祥和的浪漫情怀的环境。

沙发背景墙上的海星装饰物、水草状的壁灯；对面墙设计了壁炉，上面还有帆船、舵轮的点缀，搭配蓝白相间的地毯，整个客厅的设计仿佛将人带入了神秘的海洋中。卧室通透、明亮、宽敞、随意，主色调的蓝似乎带来海滨微微吹过的风，沁人心脾。清爽舒适的卫生间用蓝色和白色的马赛克铺就，个性十足又不失优雅。

Armstrong Ave

阿姆斯特朗大道

设计公司：Taylor Smyth Architects　建筑面积：约204平方米　摄影：Gumpesberger/Hafkenscheid

Originally constructed in 1912 as a dairy, this laneway building was converted to a residence in the mid 1980's. The exterior has been basically untouched, allowing for the delightful experience that comes from the discovery of the contemporary light filled interior when compared with the raw industrial exterior.

The new kitchen and dining room have bamboo flooring, combined with bamboo veneer on the kitchen island, creating a sense of a island which growing out of the floor. The island sits under a 4 foot x 8 foot skylight and is the symbolic centre of the house, a natural gathering spot for guests.The gypsum board walls peel back to reveal the original brick walls behind, these incidents have been used to bookend the new blackened steel gas fireplace and the surround that appears to slide out from the kitchen into the living room.

　　这个牛奶厂始于1912年，大道两旁的大厦从1980年转化成了住宅。室外的设计几乎没有什么改变，与原来的工厂外观相比把现代的灯光设计融入到了室内。

　　新的厨房和餐厅运用了竹制地板，结合岛式厨房的设计，为我们营造了一种氛围，如同这座岛就是从地板上冒出来似的。这个假岛座落在4x8英尺高的天窗下，是这个房子的中心部分，也是客人聚会的地方。石膏面板墙不经意间流露出原始的砖墙，这些不经意的设计把煤气壁炉像书立一样夹在中间，它周围的设计更是自然地从厨房过渡到了客厅。

Bahia House

巴伊亚住宅

设计师：Marcio Kogan 设计团队：Beatriz Meyer、Carolina Castroviejo、Eduardo Chalabi、Eduardo Glycerio、Gabriel Kogan、Lair Reis、Maria Cristina Motta、Mariana Simas、Oswaldo Pessano、Renata Furlanetto 项目地点：巴西萨尔瓦多
建筑面积：2165平方米 使用面积：690平方米 摄影师：Nelson Kon

This project is about a single family house which looks like a "floating house". The Bahia House makes use of the old popular knowledge that has been reinvented and incorporated throughout the history of Brazilian architecture.

These bahian houses have roofs of clay（a banal material made in a rustic manner）and wooden ceilings. The openings have large panels of wooden Mashrabiyas brought to Brazil by the Portuguese colonial architecture since the first centuries of its occupation of the American territories, and its origin is of an Arabian cultural influence. These wooden panels provide vast comfort to the interior. The traditional bahian house uses the northeastern wind blowing in from the sea to organize the floor plan and has cross ventilation in its principal spaces, always making the interior cool and airy.

The floor plan is entirely organized around a central patio, making the cross ventilation in all the spaces possible a view that looks in, to people can see a grassed garden and two exuberant mango trees. The Bahia House privileges the environmental comfort of its dwellers but does not make use of the "most modern technology" for this.

巴伊亚住宅是一个经济生态住宅,利用了巴伊亚建筑史上融会贯通的经典流行的设计理念。

拥有一个混泥土的屋顶、粗糙的方式制作的普通材质和木质的天花板。入口处有一个很大的由葡萄牙殖民建筑引进到巴西的,在一个世纪前就占领了美国市场的Mashrabiyas木质平板,而这种材质受阿拉伯文化的影响。这种木质平板为室内空间提供了极高的舒适度。传统的巴伊亚住宅利用由海面吹向建筑平面的东北风,使主要的空间变得空气流通,凉爽又清澈。

环绕着一个中心露台的平面建筑规则,使得整个空间空气流通,并且可以看见户外的人工花园以及茂盛的芒果树。巴伊亚住宅给居住者提供了一个舒适的环境,但并没有利用过多的"最现代化的科技"。

37 KKC

设计师：Takuya Tsuchida　设计公司：No.555 Architectural Design Office
项目地点：日本福州　摄影师：Torimura Koichi

This case is "A house with an Alley", single family house. The house was designed with considering the relationship to the parents house which is located at the inner part of the site. The center alley, which divides the two floating volumes, is connected to the parents house. The alley runing up and down irregularly reflects the surrounded forest geometric feature and becoms a communication space for the two families. The existing parents garden was reserved under the new resident for the two families to enjoy lunch and conversations. The residence was designed for the two families who have close relationships and happy life together.

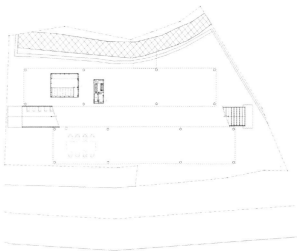

　　本案"曲径通幽",是一个独院住宅。此房的设计考虑到其与位于此住宅区内部的父母房子的关系。两个悬浮式的楼房中间有一条小径将其分隔开,由它通向父母的房子。这条反映了周围森林的几何特征的小径蜿蜒曲折地延伸着,成为这两个家庭的交流空间。原本的父母房子的花园被保留下来,供两个家庭享用中餐、沟通对话。此住宅是为关系亲密并一起快乐生活的两个家庭而设计的。

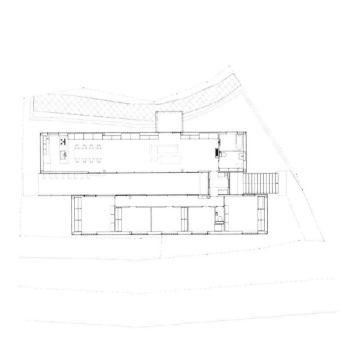

32 KRE

设计师：Takuya Tsuchida　结构设计：Alan Burden　设计公司：No.555 Architectural Design Office
项目地点：日本东京　摄影师：Torimura Koichi、Koyama Sunichi

This case looks like the floating house,is a single family house.The house is located in the high-end residential area. The most remarkable requests in the various clients' demands are "garage space for nine cards", "the most favorite car in the living room" and "high tree in the living room". Since all the requests are not able to be fit on the project site, the building itself was designed as big "Ling Room". Utilizing the entire basement, nine cars are able to be parked. Various height and size rooms are floated randomly in the big living space. Structurally, light steel structure boxes are hanged from or attached to the reinforced concrete box. These randomly floating boxes keep adequate distance and height with the car and the high tree and create attractive open space. Spaces under the "boxes" have invisible borderline and each space has the different functions according to the height. Though there are numbers of large scale items in this house, instead of setting the usual layer of floors, random floating boxes make the spacious living area possible.

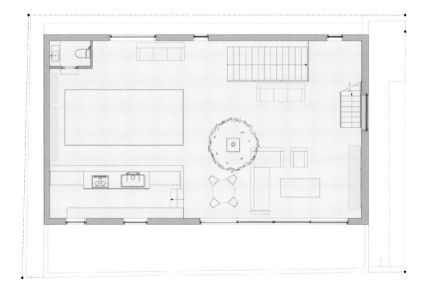

　　本案看起来像是"飘浮的房子",是一个独院住宅。此房位于高端的住宅区。客户的诸多要求中,最值得注意的是"容纳九辆小车的车库空间","客厅中要有最喜爱的汽车","客厅中要有高树"。所以设计师决定充分利用整个地下室,以便停放九辆车。宽敞的生活空间内,不同高度和大小的房间随意散布其中。结构上,轻便的钢结构"盒子"或悬挂于钢筋混凝土"盒子"或附于其上。这些随意浮动着的盒子与汽车和树保持适当的距离,营造出宽敞而有魅力的空间。"盒子"下面的空间的界限并不显眼,根据高度不同,每一个空间安排了不同的功能。虽然房子内有许多大型物体,设计师通过这些漂浮的"盒子"而不是通常的楼层,使居住空间宽敞大气。

33 Huijing Garden

汇璟花园

设计师：李仕鸿　设计公司：汕头市一帆环境艺术设计有限公司
项目地址：汕头市澄海区　建筑面积：约151平方米
主要材料：水曲柳饰面板、喷砂玻璃、镜、墙纸、浅啡网石板、人造沙安娜石板

This is a contracted European style case, the designer combines traditional European design style with modern fashionable decorative skill and retains the graceful quality with the modern feature of the age.

The design skill of the white carved flower on the TV setting wall of the living room combines with the light coffee color wall line is original. The silvery frame of the sofa setting wall is decorated with tawny mirror and it forms a strong contrast with the materials and colors. The silvery flower carved partition divides the living room and dining room. The glass window of the hallway to the children's room, kitchen, and study room is decorated with sand blasting patterns, they are like symbols throughout the entire space, transparent, dense, in diversified shapes, and they become the punchline of the space, creating a unity space as well.

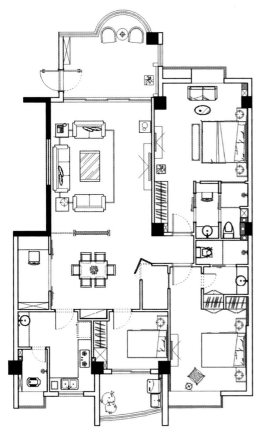

本案为简欧风格，欧式的传统元素揉合时尚现代的装饰手法，保留尊贵格调又充满时代气息。

客厅电视背景墙上面的白色刻花与下部的浅啡色墙角边的结合形式新颖。沙发墙面银色条状画框内镶茶镜，材质与色彩对比鲜明。银色雕花隔断起到划分客厅与餐厅的作用。通往童房、厨房和写字区的玻璃上都有着与客厅背景花纹一样的喷砂图案，如符号般贯穿在各个空间里，或通透、或密实、或大或小，各不相同，成为空间主旋律，使空间整体感十足。

34 Sansheng Bali Island

三盛巴厘岛

设计师：朱琦　方案审定：叶斌　设计公司：福建国广一叶建筑装饰设计工程有限公司
项目地点：福建福州　建筑面积：120平方米
主要材料：仿古砖、壁纸、黑镜、色漆、黑色马赛克

In the living room, various colors are mixed and matched with one another, which appear fashionable. The layout is minimalist yet magnificent, and the black ceiling allows the space to appear higher, and it also relieves the bad influence caused by wide area of suspend ceiling. The black mosaic and white engravings on the edge become another highlight of the living room.

At the hallway, the door is white with black frame, and the ceiling is decorated with black mirror glass, the wall on the left is decorated with many black and white paintings, and on the right, there are two black artworks. One would feel the fashionable and modern air upon entering the door.

In dealing with the space, the designer tries to use diversified methods to make a visual contrast of those elements, and connect them closely with detail dealings.

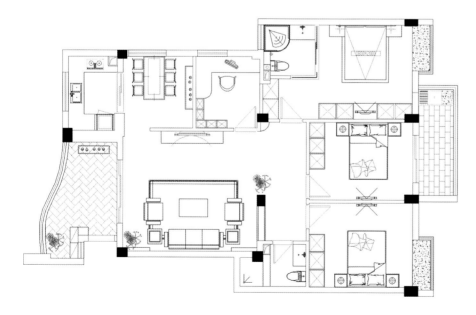

　　客厅多种颜色的搭配表现了颜色元素方面的时尚，造型简洁而不失大气，黑色的顶面更能凸显层高，缓解了大面积做吊顶带来的负面影响，黑色马赛克和边上白色雕花的搭配让客厅又多一道风景线。

　　玄关处白色黑框的门，用黑镜装饰的天花，左边是有数不清的黑白挂画装饰的墙面，右边则陈列了两尊黑色的艺术品。一进门，就让人感觉到强烈的时尚、现代气息。

　　设计师在这个空间的处理上尽可能地多元化，在视觉形成极大反差的同时，又通过一些细节的处理把它们紧紧地牵连在一起。

35 Luxury Residence by the Riverside

缇香水岸

设计师：朱琦　方案审定：叶斌　设计公司：福建国广一叶建筑装饰设计工程有限公司
项目地点：福建福州　建筑面积：120平方米
主要材料：镜面斜拼、壁纸、密度板雕花

At the entrance, the hallway protects the privacy of the bedroom and relieves the dullness of the corridor. The dining room and living room share parts of the area, which extends their visual effect. The dominated color of the living room gives people visual freshness, and the fashionable layout leaves people with striking visual impression.

Entering the space, an impressive modern life picture is shown before you, and classical elements are incorporated. A fantastic match brings an excellent living space. Minimalist lines and modern decorative materials outline the style of the living room, and the magnificent, delicate European crystal droplights promote the culture taste.

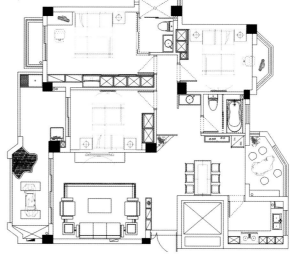

　　入户的玄关设计即解决了卧室区的私密性，又让过道不显单调。餐厅与客厅间的共用区域大大增加了空间感，让餐厅和客厅的视线得到最大延伸，客厅的主体色调让人的视觉清爽无比，时尚的造型让空间充满视觉冲击力。

　　进入空间，展现在眼前的是一幅气势不凡的现代生活画卷，其中又存在着古典元素。一种绝妙的搭配手法带来精彩的生活空间。客厅简洁的线条和现代装饰材料构成主风格，点缀豪华精致的欧式水晶吊灯，提升了文化品位。

36 Central Beautiful Garden

中央美苑

设计师：何华武、林巧英、林航英、陈亿元 方案审定：叶斌
设计公司：福建国广一叶建筑装饰设计工程有限公司 项目地点：沙县
建筑面积：125平方米 主要材料：奥立克、天伦

This project is about a duplex house, with graceful environment and fresh air surrounded it. How to lead the beautiful scenery into the interior? After careful consideration, the designer finally decided to adopt American countryside style. It gives off natural fresh air, making people feel relaxed and pleased, people would feel its easy yet abundant warmness, and the space becomes broader, brighter and lively.

White furniture and delicate fabrics are used to decorate the space, milky iron-work candlestick is with American countryside flavor, and the delicate countryside wallpaper shows the elegance and freshness of nature. Countryside dream leads people away from the noise and vanity and enables them to live a natural and easy life!

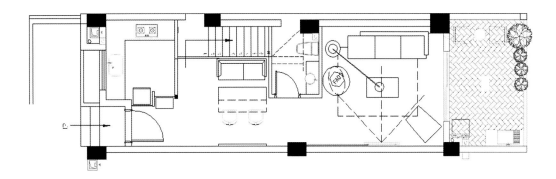

　　本案户型为复式楼，周边环境优雅空气清新。如何将这般美景复制于室内？经过设计师再三研究，最终敲定采用美式田园风格。它流露的自然清新气息，能使人感到轻松愉快，充分感受到它饱满的温情和质朴的暖意，更加开阔、明朗，极富生活情趣。

　　设计以白色家具、精巧的布艺来点缀空间，乳白色的造型铁艺烛台显露美式乡村的气息，精致的田园墙纸表露出了大自然的雅致清新。田园梦，让人们远离喧嚣，抛弃浮华，追寻最自然、最没有压力的生活！

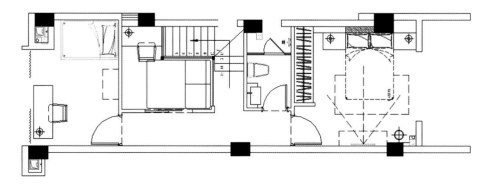

Noble Black

黑色流金

设计师： 林煜毅、陈小琪、杜绍惶　设计公司：汕头市亚太嘉毅室内设计有限公司
项目地点：金叶岛　建筑面积：220平方米
主要材料：爵士白石板、红橡饰板、深啡网石板、进口墙纸、布包、乳胶漆

In this project, the white ceiling and black plaque give people strong visual impact, the steps adopt dark coffee slate, which is noble and elegant. The entire space is mainly in black and white, expressing fashionable flavor. In the living room, the TV background wall is decorated with clear-textured volakas slab, matched with black table-board, forming color contrast and the extending of lines. The designer rejected redundant decorations, minimalist ornaments in the space demonstrate the tone of low-key luxury. Several colorful flowers make the space unique.

The background wall of the bed in master bedroom is decorated with grey velvet soft furnishings, cozy and comfortable. There are warm-colored tubes hidden under the black slabs. The designer especially designed grey patterned wallpaper in the bottom, so that the TV background wall of the master bedroom becomes minimalist yet not simple.

　　本案玄关处纯白色的天花与黑色的饰板，带给人们视觉上的冲击，台阶采用深啡网石板，高贵而典雅。整个空间用黑白的搭配诠释出时尚的韵味。客厅电视背景墙纹理清晰的爵士白石板配以黑色的台面，形成明暗的对比与线条的延伸。设计师摒弃了多余的装饰品，简练的不锈钢饰品在空间阐释了低调不乏奢华的基调。花瓶里几支颜色各异的的花，让整个空间别具一格。

　　主卧睡床背景墙以灰色的绒面布包，温馨舒适，黑色层板下藏有暖色光管。设计师特地在底部贴上了灰色带花纹的墙纸，使得主卧的电视背景墙简约而不简单。

38 Fang House

芳House

设计师：Roy.T.X　设计公司：福州柯恩罗伊景观设计有限公司
项目地点：厦门　建筑面积：260平方米　主要材料：水泥漆、金刚板、仿古砖、玻璃

This case uses black, white and grey color as its main color tone, as for the layout of the space, the designer adopts broken and combining way. The design concept of this case is to retain and borrow light from natural condition, the total area of three-storeyed building is 280 sqm wide, including 2 balconies, 3 storage rooms, and there are 38 bulbs inside the interior.

The beveled entrance retains the transparence while acting as the hallway, and the corner of bathroom makes the bathroom unclosed. The living room adopts irregular connection way, different stuffs with different materials and they are irregularly arranged, in addition, there are some red, yellow and blue ornaments as well. The transparent space makes the depth and hierarchy sense perfectly revealed in this space. The bookcase looks like a shiner, and it is the main light source of the hall and study room.

The bedroom, dressing room, bathroom are separated by two crossed transparent lines. And there is a lifting clotheshorse in the chest. It is quite necessary for the well-ordered bathroom to separate dry and wet area.

The crossed space and shiny cabinet add the unique beautiful visual sense to the space.

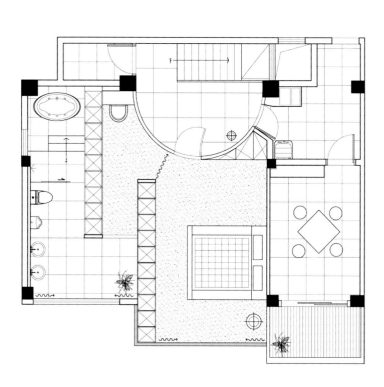

本案以黑、白、灰为主色调，空间规划上力图破格、融合。灯光设计的理念是惜光、借光。本案3层楼共计280平方米，包括2个阳台、3个储藏间，所用的灯泡总数是38个。

斜切式的入口在保持通透的基础上达到了玄关功能。卡视角的方式让洗手间不再"封闭"。客厅用不同材质的大体块不规整地交错堆叠，红黄蓝的点缀使整个空间活泼起来。书柜如同一个发光体，是大厅和书房的主要光源。书房的软木可以让你放心地把喜爱的小物件钉上去。

主卧室、更衣间、卫生间用交错的两排透明衣柜隔开。全通透的空间使纵深感、层叠感表现得淋漓尽致。

交错的空间和发光的柜体，给空间增添了独特的视觉美感。

39 Fuzhou Confucian Family

福州大儒世家

设计师： 施传峰、许娜　设计公司：福州宽北装饰设计有限公司
施工单位：福州零距离装饰设计有限公司　项目地点：福州　建筑面积：110平方米
主要材料：L&D陶瓷、大自然地板、TATA木门、欧露莎洁具、帝王洁具、美邦硅藻泥、
百姓家具、欣昇布艺窗帘、家喻灯具、德高防水涂料、多乐士漆、西门子开关、西门子家电、美森马赛

The entire design style is created by simple furnishing, furniture, and droplight, the whole design makes people feel fresh and astonished. The reddish brown leather sofa adds the luxurious feeling to the space, the simple decorative painting on the sofa setting wall is with modern sense. Beside the sofa, there is a black piano which makes the entire space full of artistic feeling.

The droplight of dining area is in the same style of the living room which looks like a blooming white flower. The black marble dining table is with strong modern sense, the slivery double fishes decoration appears flexible and lively. The modern kitchen is tidy and practical, and it is the good place for cooking.

　　客厅的整体风格是由线条精简的装修、家具及吊灯等饰物共同打造，给人以耳目一新的感觉。红棕色的皮质沙发增加了豪华大气的韵味，沙发背景墙上造型简单的装饰画则极富时代感。沙发旁边放置了一台黑色的钢琴，使得空间笼罩了一股艺术氛围。

　　餐厅区域的吊灯和客厅风格一致，像一朵散开的白花。黑色的大理石餐桌具有强烈的现代感，桌上的银色双鱼造型则显得活泼灵动。现代化厨房整洁、实用，是创造美味佳肴的好场所。

40 Gongyuandao No.1

公园道一号

设计师：叶强、简华辉　设计公司：福州宽北装饰设计有限公司　项目地点：福州
建筑面积：140平方米　主要材料：i-box家居连锁、进口壁纸、仿古砖、天然大理石、橡木修色
摄影师：施凯

The living room is contracted, refreshing and pleasing. Beige wall, light color sofa, people being there would feel totally relaxed. The black carpet seems strange in this space, but it forms a harmonious match with the other decorations, giving strong visual impact. The dining room is arranged behind the hallway, and a tawny screen serves as the separation.

The master bedroom is elegant and dignified, the large bed, contemporary droplights, without any redundant decorations. Opposite to the bed, there's a wooden cabinet in the wall.

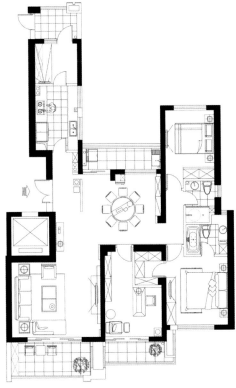

　　十分简洁的客厅，清爽宜人。米黄色的墙面，浅色调的沙发，让人置身其中，身心都能得到放松。黑色的地毯似乎与这个空间格格不入，但又与其他装饰和谐共处，给人以强烈的视觉冲击。餐厅被安排在玄关后面，由茶色的屏风作为隔断。

　　主卧雅致大气，宽大的床、现代质感的吊灯，没有多余的装饰。床对面的墙壁设置了木质的镶嵌式柜子。

Dongfang Runyuan

东方润园

设计师： 段晓东　设计公司：浙江东和东装饰设计工程有限公司
建筑面积：260平方米　主要材料：石材、玻璃、烤漆

The space is based on warm colors such as dark logwood and brown, all the furniture is specially designed, and through light and shadows to create cozy, harmonious atmosphere.

In the living room, regular solid wood and leather sofa is comfortable and magnificent. Modern flavor is created by traditional craft, matched with modest yet attractive lightings, ornaments and paintings to enhance the transparency and wholeness. The living room is contracted yet generous, delicate and with noble taste. The clean and elegant dining room is decorated with exquisite dinnerware and delicate decorative paintings, appearing minimalist yet dignified. Cream-colored tiles are used to cover the floor, which make the dining room instantly alive and light.

Coffee color is widely used in the master bedroom, which gives off deep feeling yet still be elegant. Leather soft furnishings with straight lines bed are matched with the embellishment of lightings and ornaments, one would feel as if lived in grace and fashion.

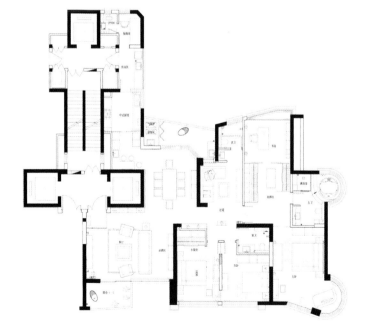

　　空间整体以深色、棕色等暖色调为主，整体设计所有的家具造型，并通过光影的变幻刻画出温馨、和谐的气氛。

　　客厅中规中矩的实木和皮结合的沙发，舒适大气。用传统工艺做出现代的感觉，加上内敛不失看点的灯具和饰品以及画的点缀，增强了通透感和整体性。客厅简约而不失大气，细腻而不失品位。干净素雅的餐厅配上精致的餐具和细腻的装饰画，既简洁又不失稳重。米色的瓷砖铺地，顿时让整个餐厅变得灵活轻盈起来。

　　主卧中咖啡色的大量使用，给人深沉的感觉又不失典雅。用皮质软包加直线条的床配上灯具和饰品的点缀，居于其中仿佛生活在优雅和时尚中。

Beautiful Residence in Jiangnan(1ˢᵗ)

江南水都美域（一）

设计师：王东生　设计公司：品步装饰设计工作室　项目地点：福建福州
建筑面积：160平方米　主要材料：水曲柳、金刚板、墙纸、仿古砖

In this home, each wall and each corner are carefully designed, which are full of vitality, bringing people different feelings. In the interior, the mild solid wood, white wall and neat geometric lines present a warm and comfortable living space.

The painting on the sofa background wall is matched with grey fabric sofa and flower or striped patterned throw pillow, forming a harmony. The corridor is clean and neat, from where one could view the living room. The staircase connecting the upper and lower floors is made of solid wood; it's designed so exquisitely, just like a beautiful melody in the interior.

Living in such an excellent environment with family members, it would be infinite pleasure to dream about future or chat while having a cup of tea.

家中的每一个墙面、每一个角落，经过设计师的精心设计，都充满活力，带给人不一样的感受。室内温润的实木、白墙，及干净的几何线条，呈现了一个温暖舒适的生活空间。

沙发背景墙的挂画，配上灰色布艺沙发，以及或印花或条纹的抱枕，使得配饰在细节上和谐。过道的处理干净整洁，从这里可望到客厅的景致。连接楼上楼下的楼梯采用实木材料，设计得如此精致，形成室内一个美的旋律。

与家人在这样的环境下，或畅想未来，或饮茶私语，都是充满无限乐趣的生活。

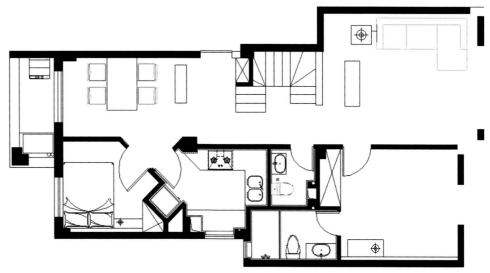

43 Dongshang Guanhu Show Flat-Elegant Life

东尚观湖样板间-精致生活

设计师：向东姝　设计公司：北京圳銮想向装饰设计有限公司　项目地点：西安
建筑面积：92平方米　主要材料：手绘涂料、壁纸、玻化砖、地板、马赛克
摄影师：崔首横

This female-flavored space is full of beauty. Color is the most important expressing technique in this case. The space is dominated with rose color, which creates a splendid and charming ambiance.

In the living room, the black end table is of primitive simplicity; the carpet of rose pattern is romantic and gentle, the designer makes the original intellectual space more delicate and sexy. The dining room is in a different style, the unique chairs and dining table create graceful dining atmosphere. The green embroidered curtain and the seating cushion of the dining chair are in the same style, they fit in well in this rose space. The entire space creates an exquisite life picture.

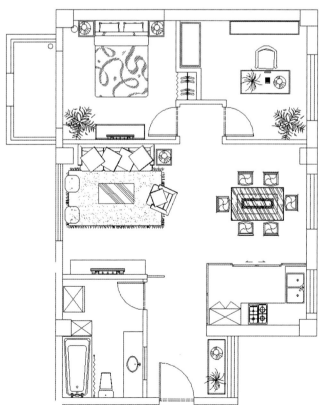

　　这个具有女性色彩的空间充满了美丽。色彩是本案中最重要的表现手法。以玫红色为主色调的空间，渲染了绚丽、妩媚的气氛。

　　客厅黑色的中式茶几，古朴自然；玫瑰花地毯，浪漫柔情，设计师用独特细腻的手法将原本知性的空间打造得如此精致性感。餐厅空间则有另一种风情，造型特别的椅子与餐桌营造了优雅的用餐氛围。绿色的绣花窗帘与餐椅坐垫风格一致，在一片玫红色中找到属于自己的定位。整个空间打造了一幅精致的生活画面。

New Dongshang Show Flat-Gentlemen's Choice

新东尚样板间-绅士主张

设计师：向东姝　设计公司：北京銮想向装饰设计有限公司　项目地点：西安
建筑面积：132平方米　主要材料：石材、壁纸、玻化砖、地板、樱桃木贴面　摄影师：崔首横

In this case, coffee is the dominated color, deep coffee furniture is matched with light coffee wall, which allows the entire space to be generous and elegant, creating a low-key luxurious atmosphere.

The white and golden throw pillows become more attractive in the coffee ambiance, giving people strong impact on visual and other senses. The two paintings of sailing boats on the sofa background wall express the owner' good wish of "everything goes smoothly." The whole wall is installed with French window, which makes the space to be clear and bright.

In the bedroom, the wide use of golden color fully manifests its nobleness, leaving people with deep impression. Even in the bathroom, golden tiles are used, appearing luxurious and magnificent.

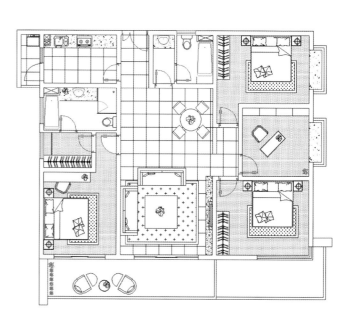

本案以咖啡色为主色调，深咖啡色的家具搭配浅咖啡色的墙壁，使整个空间看起来简约大方但又不失高雅，营造了一种低调奢华的氛围。

白色、金色的抱枕在一片咖啡色中显得特别夺目，给人视觉和感官上的强力冲击。沙发背景墙上的两幅帆船挂画，表达了主人期望"一帆风顺"的美好寓意。整面墙的落地窗让空间显得明净敞亮。

卧室泛金的色调将它的高贵气质展露无遗，让人只看一眼便会记住。就连卫浴也是采用金黄色的瓷砖，豪华大气。

Jindi International Garden

金地国际花园

设 计 师：林卫平　设计公司：宁波西泽装饰设计工程有限公司　项目地点：浙江宁波
建筑面积：180平方米　主要材料：大理石、马赛克、墙纸

"Various forms, united spirit" is the major feature of this Neo-classical style show flat. While focusing on the decorative effects, modern techniques and contemporary materials are used to bring classical flavor, and allow the space to be both modern and classical. The perfect combination also enables the owner to have both physical enjoyment and spirit comfort. Simple techniques, modern materials and technology are adopted to achieve traditional outline and feature. White and dark red are the major colors of this project; and white allows the house to be brighter. The hard white lines of the wall and delicate, elegant wallpaper are matched up with magnificent classical furniture, which is impressive and tasteful.

In this case, modern elements and traditional ones are combined so as to create an elegant and comfortable living vibe.

"形散神聚"是本套新古典风格样板房的主要特点。在注重装饰效果的同时，用现代的手法和材质还原古典气质，使之具备古典与现代的双重审美效果，完美的结合也让业主既享受到物质文明也得到精神上的慰藉。用简化的手法、现代的材料和加工技术追求传统式样的大致轮廓特点。白色、暗红色是本案的主色调，白色基调使整个房子的色彩愈加明亮。墙面硬朗的白色线条走边，细腻素雅的壁纸，搭配古典华丽的家具，稳重又不失品位。

本案将现代元素和传统元素结合在一起，营造了一种优雅而舒适的生活氛围。

FuXi Dadao

福熙大道

设 计 师：崔静　设计公司：尚层装饰（北京）有限公司
项目地点：北京　建筑面积：250平方米　主要材料：布艺、壁纸、大理石

In this project, the decorative style is Neo-classism, which is minimalist and passionate, refreshing and light, breaking away from the classical frame. The basic color of this space is white, light blue and beige are used for embellishment. As for the decoration, the aesthetic vase, lively bonsai and the contemporary decorative design echo with each other, appearing more pleasing. After being simplified of the classical complicated decoration, and combined with modern materials, the furniture presents a new look of minimalist classicism.

本案装修风格定位于新古典主义，打破了古典主义的框架，简约而充满激情，清新并不厚重。空间以白色为主色调，辅以淡蓝和米黄色，清新典雅之风扑面而来。在装饰上，颇具艺术感的花瓶、生机盎然的盆栽、充满现代感的装饰设计相映成趣。本案的家具将古典的繁复雕饰经过简化，与现代的材质相结合，呈现出古典而简约的新风貌。

47 Fanhai International

泛海国际

设计师：顾程　设计公司：尚层装饰（北京）有限公司　项目地点：北京市朝阳区
建筑面积：300平方米　主要材料：实木、玻璃、布艺

In this graceful and quiet space, one could feel light classical flavor. The spacious living room is full of complex elements, brown and white sofa is matched with the blue flower-patterned chairs beside, appearing magnificent and fulfilled. Nearby the window, the cany furniture, dark wallpaper, solid wood as well as fabric decorations are mixed in the space, presenting a visual feast for us.

The design isn't limited to fixed rules, fresh and pleasing decorations present an incomparable style. The wide area of window of the interior brings in abundant lights, allowing the space to be bright and transparent.

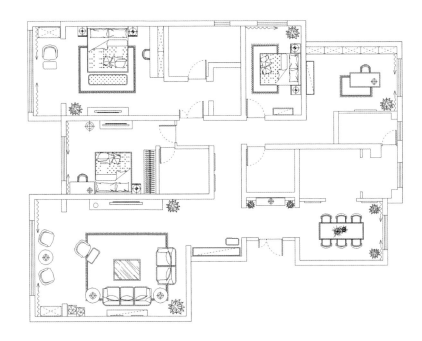

在这个优雅、宁静的空间中流淌着淡淡的古典气息。开阔的客厅充斥着繁复的元素，褐、白的素色沙发，旁边搭配了一张蓝色印花的椅子，更显华丽、饱满。窗边藤制的家具、暗纹的壁纸、实木和布艺的搭配等，这些不同质感的材料拼合在一个空间中，为我们呈现出一场视觉盛宴。

设计不限于条框，清新宜人的装饰带出室内无可比拟的风格。室内大面积的窗户引进良好的采光，使得空间更加明亮清透。

48 Natural Beauty

丽景天成

设计师：黄耀国　设计公司：福州好日子装饰设计工程有限公司　项目地点：福州
建筑面积：120 平方米　主要材料：大理石、防古砖、马赛克、明镜、墙纸、水泥漆

The entire space in this project is designed to be minimalist yet generous. It inherits post-modern minimalist fashion elements, and marble and mirror glass are used to decorate the space, which has injected new life to the space, elegant and relaxing.

The golden flower-patterned wallpaper on the sofa background wall is with distinctive modern feeling and the cabinets with 9 lattices offers the owner a convenient place for ornaments. The grey marble TV wall becomes the separation of the living room and dining room, forming a semi-open space, and one can see the living room while having dinner. The wall of the dining room is made of glass, which allows the space to appear much larger than it really is. The gauze curtain also makes it brighter and more spacious. Silver mosaic is used at the entrance of the washroom, which adds some fashionable flavor to the space.

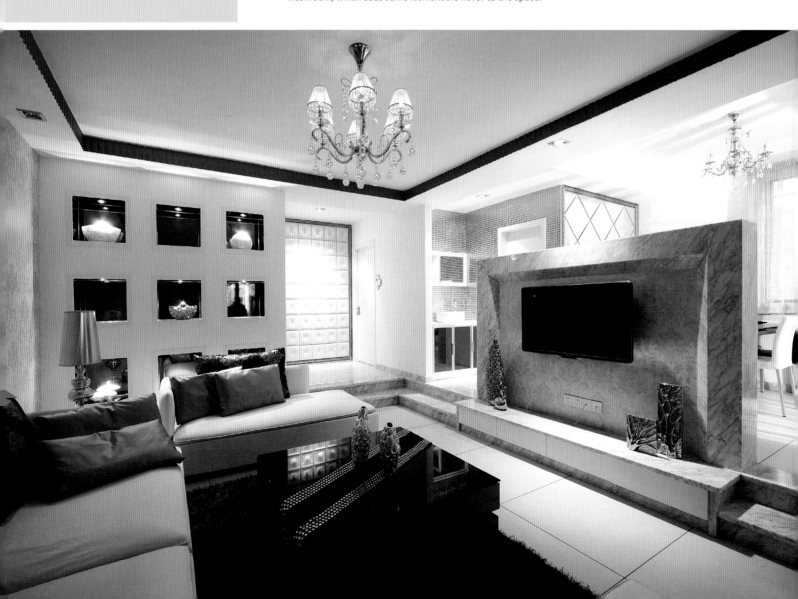

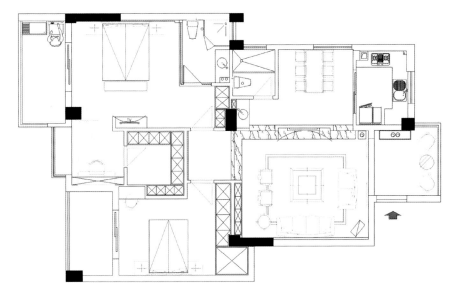

　　本案设计整体空间简洁大方，传承后现代简约时尚元素，采用大理石及镜面材质渲染空间，活跃了整体效果，典雅且轻松。

　　沙发墙上金色印花图案的墙纸富有强烈的现代质感，旁边的九宫格收纳柜方便业主摆放一些装饰品。灰色大理石的电视墙，正好也成了客厅和餐厅的隔断，形成了一个半开放的空间，人们在用餐的时候也能关注客厅的一切。以整面明镜砌成的餐厅墙壁，发挥了"一加一大于二"的空间思维。纱质的窗帘，使空间更加明亮宽敞。而卫生间门口区使用了银色的马赛克，这又为空间增添了时尚的色彩。

49 Color Charm

色韵

设计师：董龙工作室设计团队　设计公司：DOLONG董龙设计
项目地点：江苏南京　建筑面积：90平方米　主要材料：木雕、贝壳水晶帘、马赛克、进口墙纸、乳胶漆等　摄影师：金啸文

A house which is full of fairy tale flavor must be many girls' ideal home.

In this space, bright colors are widely used. The background wall of the sofa and TV set is in emerald, while in one corner of the dining room, red is used. As the wall adopts bright colors, so the furniture and floor adopt light color to contrast with them and achieve visual balance.

Due to the bright colors, there aren't many decorations in the house. Living here, you would feel as if lived in a deep forest, which is imbued with fresh air of nature. Lively bright emerald seems to tell you the coming of spring.

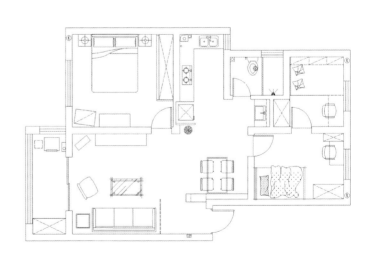

　　充满童话色彩的小屋，相信是很多女孩子梦寐以求的理想之家。

　　空间大胆地运用了鲜明的色彩，以翠绿的颜色为沙发和电视的背景墙；而餐厅更是采用了一抹艳丽的红色，作为餐厅一角的点缀。由于墙面的用色大胆，所以家具和地板都特意运用了浅色来区分，以达到色调和谐。

　　由于用色较为大胆，所以房子没有过多的装饰品。住在这里，你会感到如同身处森林深处，充满了大自然的清新气息。活泼亮丽的青翠，仿佛在悄悄地告诉你春天的到来。

50 Haiyu Xiwan

海语西湾

设计师：王五平　设计公司：深圳五平设计机构　项目地点：深圳　建筑面积：130平方米
主要材料：灰镜磨花、红橡木油白色、墙纸、抛光砖、灰木纹大理石、马赛克

For the style of the home interior, the owner has clear goals. He prefers minimalist, quiet, fashionable yet with the classical colors of black, white and grey. All the luxurious and classical elements are faded here, what left is the pureness, a symphony created by black, white and grey.

In this case, the simplest elements of black and white, virtualness and solidness, block and lines are used to express the space. Instead of using colorful elements, two contrasting colors of black and white are widely used. Besides, more artistic method is applied to create a distinctive space atmosphere.

The wallpaper with thick lines on the sofa background wall injects lively rhythm to the space of plain color. One could rest and appreciate this black and white dominated space, various other colors are used for embellishment, and the entire space appears clean and neat. In the meantime, ornaments are skillfully used to reduce the dullness caused by too much black and white.

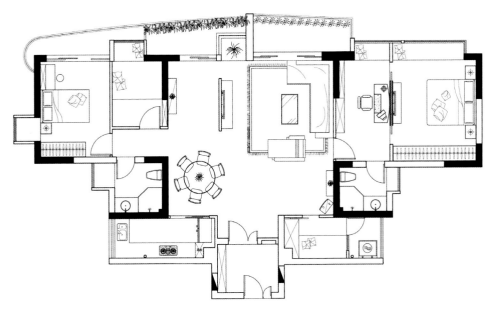

　　对于家的格调,业主有着明确的方向,喜欢简洁安静,时尚并有着经典的黑白灰。一切关于奢华与古典的畅想在这里已渐行渐远,留下的便是纯净,一场黑白灰的时尚交响曲。

　　本案利用黑白虚实、块面和线条这些最简单而有力的语言来设计整个空间,没有运用太多的色彩而是控制在黑、白两色的对比中,并在空间运用更艺术化的表现手法来营造其与众不同的空间感受。

　　沙发背景墙上大线型墙纸的运用让素色空间有着律动的节奏,在黑白的极致中静心品味,丰富的色彩在其中流淌,整个空间显得非常干净。同时,巧妙地运用饰品的搭配,避免了黑白的单调造成的空洞。

51 Zhuoyue Weigang

卓越维港

设计师：王五平　设计公司：深圳五平设计机构　项目地点：深圳　建筑面积：230平方米
主要材料：多乐士乳胶漆、抛光砖、墙纸、水晶灯等

In a serene space, one must have his own beautiful mood. Here, each detail is telling a happy story, there's no complicated decoration, but all is natural.

About the design technique, in this project, the emphasis is put on the entire space effect, without redundant decorative elements, just to show a simple life attitude and the love for life. So, though appearing simple, the elements in each space play an important role.

In this project, the most obvious change is that the kitchen and dining room are combined together. And a sliding door is used to separate them with the living room, which protects the necessary function and also beautifies the space.

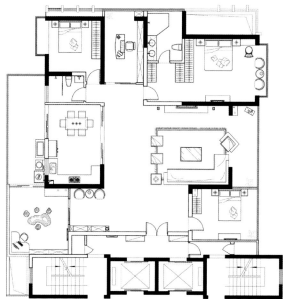

　　静谧的空间,有着属于自己的一份美丽心情。在这里的每一个设计细节语言,都畅想着一段幸福的故事,没有表面繁琐的装饰语言,但总是表现在不经意间。

　　在设计手法上,本案更注意空间的整体,没有过多的装饰元素,只为表达一种简单的生活态度和一份对生活并不简单的热爱。所以看似简洁,其实,每一个空间存在的元素,都起着极其重要的作用。

　　本案在空间处理最大的改动就是把原厨房和餐厅合为一体,与客厅交界的地方用装饰推拉门隔开,既保证必要的功能性,同时又起到美饰空间的效果。

52 House besides Hills

依山郡

设计师：吕宏伟　设计公司：深度设计顾问（香港）有限公司　项目地点：深圳市龙岗区
建筑面积：180平方米　主要材料：大理石、直纹水曲柳、镜钢、墙纸、浮雕木地板

This case adopts European pastoral style and continues the calico style. In order to keep the integrity of the whole space, the designer extends a raw of cabinets to the end and subtly deals with the relationship of two doors and living room, combining the door and cabinet together.

Another highlight of this case is TV setting wall, the entire blue color of the TV setting wall makes people feel fresh and comfortable, the elegant color gives us a sense of romantic livable feeling. The calico fabric sofa and contracted tea table give us comfortable and free feeling. The lively color makes people having a sense of brightness.

The wall of master room uses green color which looks like the color of apple, making people feel relaxed and relieved from the fatigue of the whole day.

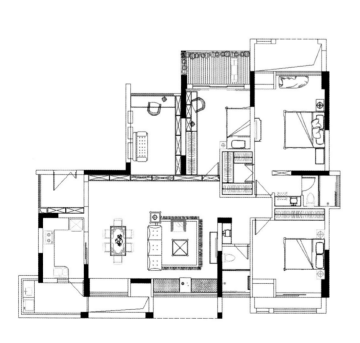

　　本案的风格为欧式田园风，延续了小碎花的情缘。为了不破坏整体性，设计师用一长排的柜子延伸到头，巧妙处理了客厅的两扇门与客厅的关系，把房门与"柜"结合起来。

　　本方案还有一亮点就是电视背景墙，整面墙的蓝色让人感觉清新怡人，淡雅的色彩散发出浪漫的生活情调。碎花布艺沙发和极简洁的茶几带来舒适性和随意性。活泼的色彩，则给人明朗的感觉。

　　主卧的墙面采用如苹果般清脆的绿色，使人不由得身心放松，顷刻间纾解了一天的疲劳。

53 Xiangrui Garden Huo's Villa

香瑞园霍公馆别墅

设计师：张逸群　设计公司：深圳居众装饰　项目地点：深圳　建筑面积：500平方米
主要材料：蚀花茶镜，黑色镜面不锈钢、古木纹大理石、遍地黄金大理石、爵士白大理石、橡木、进口马赛克、不锈钢、进口布艺、水晶扣、进口墙纸

The project is to design a luxurious house for "dancers". Luxurious, splendid and fashionable are the design style, showing strong personal flavor.
The designer uses different materials and cold colors to create a cool, charming space, showing fashionable quality. The space construction, color and lighting are matched so well that they complement each other to show the theme-charm. The match of fabrics, leather and crystals is extravagant and attractive. The distinctive furniture, transparent droplight as well as the varying lighting make people feel like being in a splendid stage, so fascinating.

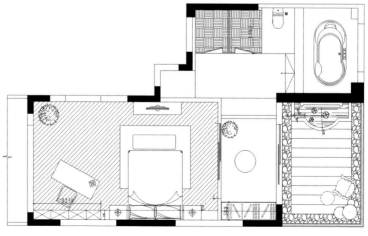

　　这个案例是要建造一个为"舞者"量身定做的豪华府第。奢华、艳丽、时尚是整体的设计风格，表达出强烈的个人色彩。

　　设计师用不同的材料及冷酷色调构成炫酷、另类的空间格调，展现了时尚的气派。空间构造与色彩及灯光的搭配相映成趣，力求体现本案的主题——魅。布艺、皮革与水晶的搭配是绝对的奢靡，而且撩人心弦。造型别致的家私、晶莹剔透的水晶吊灯，变幻多端的灯光处理，使人仿佛身处于华丽舞台之中，纸醉金迷。

54 Sky-blue G-type Duplex Villa

蓝郡海蓝G型复式豪宅

设计师：张逾群　设计公司：深圳居众装饰　项目地点：深圳盐田　建筑面积：210平方米
主要材料：蚀花茶镜、爵士白大理石、橡木、进口马赛克、不锈钢、进口布艺

The concept of this villa is using silver-white as main color and a large number of mirrors to enhance the flexibility and luxury of the space, so that the masterpiece of the perfect match of different elements is reflected from the delicate style home interior.

The soft color of the wall endows the space with peaceful and comfortable vibe. The most pleasant essence of the design lies to the transparent glass between the living room and the stairs, which is just like a magic painting, adding more beauty and practicability to the elegant living room. The color and the tactile material form a perfect match, enriching one's visual senses. The quite living room is full of romance. Independent place and soft light make people enjoy the life quality and the comfort of home.

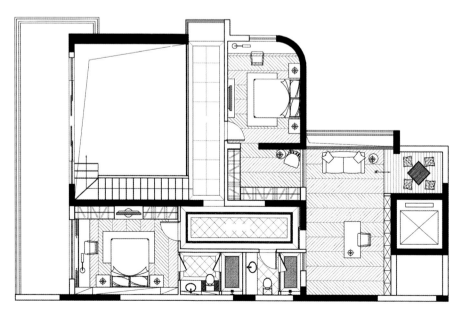

　　本案的设计概念以银白色为基调，加以大量的镜面搭配，提升空间的灵动与奢华，使居家在细腻精致的风格中透出不同素材搭配而成的杰作。

　　柔和的墙面色调，体现出整个空间的宁静舒逸。以无暇通透的玻璃作为客厅与楼道之间的隔断，就是精髓所在，犹如一幅可"魔术"的艺术画。雅致的居室增添美感和实用性。色系与质感的交融，丰富了眼目感觉。恬静的居室充满绮丽的情调，一处处独立的空间与柔和的点光源，霎时让人觉悟生活的品质及居家的安逸。

55 Qianlong Manor

钱隆首府

设计师：陈晓丹　设计公司：福州佐泽装饰工程有限公司　建筑面积：128平方米
主要材料：仿古地砖、木纹石、防腐木、实木地板、钢化玻璃、整体厨房、艺术墙纸

In this project, the house includes three bedrooms, two living rooms and a study. At the entrance, there is a corridor-alike little garden, which contains two vertical levels, appearing still and mobile, which is the distinctive feature of this structure. As for the design, the original structure has been fully utilized, minimalist and clear lines are adopted to divide the space reasonably. The open style living room, intercrossed and tidy modeling and steel clear glass railing bring transparent feeling, which eliminates visual depression and allows the tired owner after a day's work to feel relieved. Minimalism means a lot-with few decorations and sufficient functions, it meets the requirement of contemporary people who pursue for a simple life.

　　本案为三室加一个小书房两厅的居室设计，长廊式的入户小花园，上下错层，动静分明，成为这个结构的鲜明特点。在设计上充分利用了本身结构的优势，采用简约明朗的线条，将空间进行了合理的分隔。开放式的大厅、错落有致的造型设计及钢化清玻璃的栏杆扶手给人予通透之感，避免视觉给人带来的压迫感，可缓解业主工作一天的疲惫。少即是多——装饰少，功能多，这正满足了现代人渴求简单生活的心理。

56 Vanke Jinse Chengpin

万科金色城品

设计师：韩松　设计公司：深圳市昊泽空间设计有限公司　项目地点：广州
建筑面积：86平方米　主要材料：大理石、地毯、皮革、水晶灯

The broad space shows its ambience and details as well. A huge mirror surface on the sofa background wall extends from the ground floor to the first floor, matched with the crystal droplights on the ceiling, making the living room more resplendent. Fabric sofa, end table and various adornments add the finishing touches.

The second guest-receiving area is in modern fashionable style, and it becomes the video center of the home. It adds fresh fashionable flavor to the space, forming a contrast with the main guest-receiving area.

The entire space leaves people with the impression of massiveness, nobleness, elegance and simplicity.

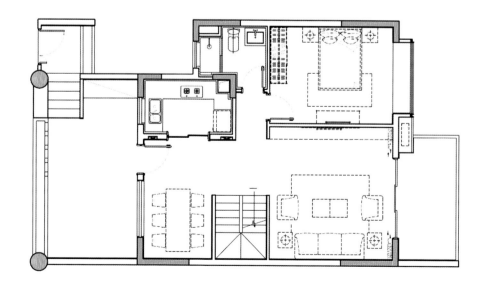

　　宽大的空间既显示出了气魄，也显示出了细节。沙发背景墙上巨大的镜面，从一楼一直延伸至二楼，加上天花的水晶吊灯，将客厅装饰得更加金碧辉煌。布艺沙发、茶几以及各种小配饰起画龙点睛的作用。

　　第二会客区以现代时尚的风格出现，形成家庭影视中心，与主会客区在风格上形成对比，掺进了一股清新的时尚感。

　　整个空间给人的感觉敦厚稳重、端庄气派、含蓄优雅、淳朴古拙。

57 Starry Lake Scenery Duplex Show Flat

星湖丽景复式样板房

设计师：陈琦、周泽军　设计公司：汕头市丽景装饰设计有限公司　项目地点：汕头
建筑面积：174平方米　主要材料：花贝马赛克、灰镜、橡木白漆、韩国墙纸、欧式丝绒

This project belongs to small type duplex houses, and it adopts light color countryside style.

On the first floor, the kitchen and bathroom were originally connected and faced the living room, the designer designed a hidden door at bathroom, which protects its privacy and makes the living room more ordered and beautiful. As the house is facing south, the designer used transparent dividing method to let the air flow smoothly. If you sit in the living room and have a rest, you would feel the gentle breeze on your face. The dining room can also be seen in the living room, so the delicate small house appears more spacious.

The study on the second floor is the room where the owner work or have fun. Clear glass with European velvet is used as partition, which extends the European countryside theme and also serves as a beautiful ornament. The bay windows in the two rooms are redesigned to be cupboards, which satisfied the owner's need of storage.

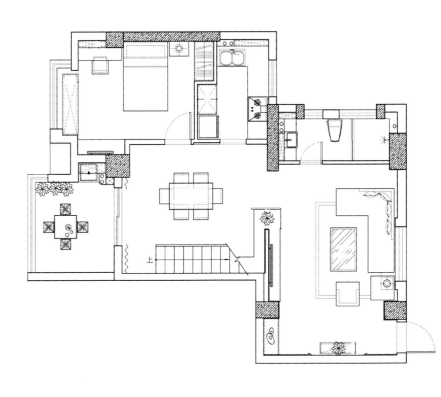

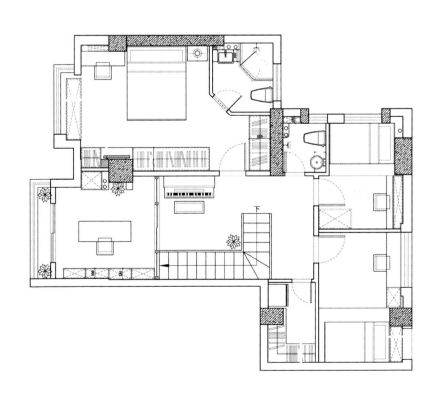

　　本案为小户型的复式单元，采用浅色调的田园风格。

　　一层原建筑结构为厨房与卫生间合在一起，面向客厅，设计师在卫生间设置了隐藏门，既保证了卫生间的私密性，又使客厅完整而美观。由于房子坐北朝南，设计师用通透隔断的手法，让空间南北对流。坐在客厅休闲的时候，能够感受到微风拂面。餐厅的景观纳入客厅中，使这个精致小户型更具空间感。

　　二层的书房是业主日常工作休闲的地方。采用清玻璃夹欧式丝绒作为隔断，既延引了欧式田园的主题，又让这个隔断成为一个漂亮的装饰。两个房间的飘窗都改装成储物柜，满足了业主的储藏需求。

TOP TEN CHINESE DESIGNERS

中国十大当红设计师

2010~2011年首届"中国十大当红设计师"评选活动圆满结束。本次大赛是由《室内公共空间》杂志社、深圳市创扬文化传播有限公司主办,大连理工大学出版社、华中科技大学出版社、福建科学技术出版社、上海科学技术出版社支持。本刊与出版社合作,以最受读者欢迎为宗旨,评选出十位其作品最受消费者青睐的当红设计师。

邱德光 (T.K.Chu),出生于中国台北,亚洲著名设计师,淡江大学建筑系毕业。在建筑与室内设计业界超过三十年经验,参与了两岸许多高级室内设计案与建筑公共区域的规划案,如著名的台北信义之星以及北京星河湾等众多著名寓所。熟悉运用华丽、艺术、时尚元素,将生活形态和美学意识转化为富贵身份的设计大师。在高级名人住宅、饭店业界及商业空间上,颇受赞誉。目前被媒体誉称为国际ArtDeco装饰主义大师。
邱德光以他多年深厚的美学素养利用具有时尚装饰作用的元素结合当代空间设计,赋予生活新奢华内涵,成功塑造当代东方都会美学。

利旭恒,出生于中国台北,英国伦敦艺术大学 BA (Hons) 荣誉学士,Golucci Design Limited 古鲁奇公司设计总监,长年致力于酒店餐饮商业空间与商业地产的室内设计工作,多年的酒店餐饮项目设计过程,累积了丰富的项目经验,在多个设计风格与商业型态上都卓有成就,2010年《domus China》杂志评为 60 China Interior Designers,更成为国内知名的餐饮空间室内设计师之一。
利旭恒的作品多次收录在两岸媒体杂志上,如《Interior Design》(美国室内)、《id+c》杂志、《FRAME China》、《Interior 台湾室内》、《缤纷Space》杂志、《DOMUS》、《现代装饰》、《设计家》杂志等。
近年获奖情况:
《Interior Design》(美国室内)2010年度金外滩最佳材料运用优秀设计奖
《Interior Design》(美国室内)2010年度金外滩最佳餐厨空间优秀设计奖
《Interior Design》(美国室内)2009年度金外滩最佳餐厨空间优秀设计奖
中国建筑装饰协会2009年中国室内空间环境艺术设计大赛餐饮空间设计奖
中国建筑装饰协会2008年度中国十大样板房间设计师
《豪饰》杂志2008年度中国第二届十佳配饰设计师
City Weekend 2008北京最佳火锅餐厅称号(作品《鼎鼎香餐厅》)
That's Beijing 2007北京最佳餐厅称号(作品《鼎鼎香餐厅》)
中国建筑装饰协会2006年度中国十佳设计师
中国住文会2004年度中国十佳设计师

刘伟婷,近年被The Andrew Martin International Awards选为全球著名室内设计师之一,而当中的亚洲女设计师可谓少之又少。她于2004年创办刘伟婷设计师有限公司,后又于2007年成立麦麦廊(专业配饰顾问及销售服务)。拥有多年丰富设计经验的她,凭着多元化的素材、新颖意念、敏锐触觉,发挥形形色色的创作。公司成立至今已完成超过200个项目。
近年获奖情况:
2009~2010年度中国国际设计艺术博览会"资深设计师"及"中国设计艺术成就人物"荣誉
中国建筑装饰协会设计委员会以及《家饰》杂志"2009年度中国十佳配饰设计师"
2008~2009年度中国国际设计艺术博览会"最具影响力优秀设计机构"及"设计银奖"
《创意中国》近年内地、港、澳顶尖室内设计师前三十名中唯一女性。

林文格,L&A文格空间设计顾问(深圳)公司创始人,高级室内建筑师、国家杰出室内建筑师。现任IFDA国际室内装饰设计协会理事、ICAD国际A级职业景观设计师、中国建筑学会室内设计分会(第三)专业委员会委员、香港室内设计协会中国深圳代表处委员中国建筑学会会刊《a+a》顾问委员会委员。意大利米兰理工设计学院室内设计硕士。
近年获奖情况:
2009年美国Hospitality Design Awards酒店空间设计大赛最高荣誉winner大奖
2009年世界酒店"五洲钻石奖"——"最佳设计师"
APIDA第十一届亚太区室内设计大奖餐馆酒吧类别冠军奖
APIDA第十四届亚太区室内设计大奖酒店类别铜奖
2008年"华耐杯"商业方案类二等奖
2007年"金外滩"提名奖
2007年"金外滩"最佳餐饮空间奖
2006年INTERIOR DESIGN 酒店设计奖
2006年中国(深圳)室内设计文化节城市荣誉奖
2006年中国(深圳)室内设计文化节酒店设计类别金奖
2006年第三届海峡两岸四地室内设计大奖赛五项大奖
2006年《现代装饰》(国际)室内设计传媒奖年度精英设计师
2006年中国建筑学会"全国百名优秀室内建筑师"
2004年中国建筑学会"全国百名优秀室内建筑师"

邱德光　利旭恒　刘伟婷　林文格　陈明晨